Green Eyes & Black Rifles
Warrior's Guide to the Combat Carbine

By Kyle E. Lamb

Edited by Kelsey E. Lamb

© 2008 Viking Tactics

Trample & Hurdle Publishers

Third Edition Rev.

WARNING

The practices, methods, and techniques described in this book are not for beginners. This book is designed for persons with background, knowledge, and experience in firearms and firearm handling. If you do not have substantial background, knowledge, and experience with firearms and firearm handling, seek instruction before attempting to utilize or implement the practices, methods, and techniques described in this book.

Shooting and handling of firearms involves inherent and unavoidable risks. Firearms are dangerous. If you are not willing and able to take responsibility for your actions, do not handle firearms. Whenever handling a firearm, remove the magazine (if applicable) and visually check the chamber before handling, cleaning or working on the firearm to verify that it is not loaded. Always use appropriate eye and ear protection whenever shooting. Know how to fire and otherwise operate any firearm you are handling. Refer to the applicable owner's manual for specific instructions. Many firearms are different, even if they appear to be similar. Your firearm may not operate in the same manner as the firearms described in this book. Never handle a firearm except in an appropriate location, such as a shooting range.

Always observe the four (4) primary rules of gun safety:

1. **Treat all firearms as if they were loaded.**

2. **Never point a firearm at anyone or anything you are not willing to put a bullet through.**

3. **Keep your finger off the trigger and out of the trigger guard until you are ready to fire.**

4. **Know your target and what is behind it.**

This book is not intended to be a complete or exclusive guide to firearm operation, and it particularly does not purport to give complete instructions with regard to any particular firearm. This book must be used in conjunction with adequate prior experience and other resources.

Disclaimer: *No one associated with this book, including its author, is or will be responsible for death, injury, loss or damage to property due to the use or misuse of firearms.*

SHOOT SAFE!

Dedicated to all the Fallen Eagles

ACKNOWLEDGEMENTS

There have been a ton of folks that have helped me to see this from start to finished product.

First and foremost, my family. Melynda has put up with long hours of my hacking away on the computer, saying that someday there would be a book. My son Lukas had to hold down the ranch and be the man of the house while I was at war.

Shooters, special thanks to the men that showed me which end of the rifle I was supposed to point at the bad guys. Ray Bob Gentry, Matt Rierson, Brian McKibben, John Cone, Bennie Cooley, Mike Voigt, Rob Leatham, Jerry Miculek, and Jim Clark.

My daughter Kelsey who took over Editing duties after Tom Badgett had to bow out due to his wife's health issues. Kelsey has the guts to tell me it is stupid, if it is in fact stupid. She simply said "Dad do you really want to say this?", and then she fixed it so it really made sense.

To my friends. Those that have promised to buy a book, just to humor me.

Mike Pannone for his in-depth understanding of the weapon system and dealing with malfunctions.

Ray Comley for the photo work. A fellow 5th Special Forces Group soldier who hung out in the sun at his farm for several days to get the pictures we needed.

Trey L. and John P., two Warriors who are in the fight as we speak. They took the time to help me with photos of the real deal.

Kevin Talbot and Floyd Dean, they have taken pity on my lack of artistic talent and saved me from embarrassment.

Matt Frederick, responsible for book layout and all around fire extinguishment responsibilities.

Dave Vandenberg and John Waters, they have kept me out of jail, somehow?

Tom Badgett, the originator of the book idea. Tom participated in one of the classes we conducted in Tennessee. He told me he was an Editor and would help me if needed. He tricked me into not only getting a decent outline, but starting the book project. It has taken 5 years and I am grateful for Tom's patience as I asked stupid questions. I have heard there are no stupid questions, I think Tom may disagree.

Lastly, thanks to the Warriors who helped me get where I am today. The Warriors who will continue the fight, and the Warriors who will build their skills and test their temper against enemies, foreign and domestic, now and during the upcoming fight.

Kyle E. Lamb

INTRODUCTION

I've been exposed to numerous tactical instructors and hundreds of shooting scenarios. I've faced real world confrontations involving the exchange of small arms fire. I've worked with some of the best-trained fighters in the world, and some who only thought they were. I can tell you first hand that the right tactical training and warrior knowledge spell the difference between walking away from a firefight and being carried away.

The successful fighter never stops learning. You must attain knowledge anywhere you can find it; you keep what works and you throw away what doesn't. You keep yourself prepared for that one fateful day when your life and the lives of others with you depend on how well you've learned your craft. You must attain Knowledge Dominance in the tactical arena; learn, learn, and then learn some more.

No one instructor, no one reference, no one experience can teach you everything you need to know. That's why I decided to write this book, to add whatever I can to the tactical shooting body of knowledge, hopefully to give even one person just the right edge at the right time to make a difference.

Now I'll tell you up front, I am not a writer. I am not a gunsmith. I am not a philosopher, poet, or English major.

I am a tactical shooter.

My experiences have made me what I am today. I have been very lucky to meet the people that I have. These great men have given me tons of tactical knowledge to consume, rehearse, train, test, modify, and employ. Once I have practiced and refined these techniques, I try to pass on everything I've learned to others in my chosen profession. I have also been heavily involved in competition shooting with pistols, rifles, carbines, and shotguns. This competition experience has helped to make me a better-rounded tactical shooter.

Competition is a great place to experience shooting stresses that you might otherwise not experience until the day you become engaged. I, for one, want to feel the stress in competition, know what it is like, deal with it, and learn from it. Being overcome by events in a tactical crisis is not the time to feel your first dump of adrenaline. So get out there and place yourself in an unfamiliar, stressful environment. It will pay off.

This book, however, is not for the competition minded; this book is for the shooter who hopes to use ballistic tools to eliminate a threat if the need arises. You may be a law enforcement professional, full-time military, National Guard or Reservist, tactical consultant, or instructor. You will find information here that you can use along with practical training techniques and exercises to improve your skill.

My intent is to give you a tactical reference book that you can use for years to come. I also intend to deflate several "Urban Legends" from the tactical community. I can say 'I have been there and this worked, or did not work for me.' And as anyone who is successful in their field must evolve, we as tactical shooters must evolve. The AR rifle has evolved, our ammunition has evolved, and we must evolve to better tactics and techniques as they become available.

Try different techniques and equipment. If it works for you, add it to your Tactical Toolbox. If it doesn't, add it to your Toolbox anyway and try it again later. You may need to train with the basic fundamentals for awhile before you become comfortable with some of the more advanced techniques I will discuss.

This book will focus primarily on the AR rifle and its variants. If you have chosen or been assigned a different weapons system you still will find the material useful. Most of the shooting positions I discuss are not specific to any particular weapon system. You should be able to apply my techniques to your rifle, shotgun, or sub gun.

When I instruct I usually have two types of students, those who are assigned their weapon, and those who purchased their weapon with their own hard earned cash. Either way, my goal is to instruct them to be the best that they can be with what they have. If a student shows up with a wooden club, we will make it work. If we have to drive nails through that club to make it more effective, that is what we will do.

As you study this book, as you practice the techniques I discuss, as you work with any tactical information, keep an open mind! I have trained with instructors who believe the AR is a poorly designed weapon and that the ammo will not produce the down range results you will need in a tactical engagement. You probably could say the same about virtually any weapon system, and you'd be right some of the time. (You can drive a nail with a screwdriver, but it isn't the best choice of tool for that job.) But overall, within the context of the tactical situations for which this book is designed, the AR is an awesome rifle, with awesome potential. Its modularity has made it one of the most widely used tactical firearms available today. Learn what works; you will not be let down when the time comes.

I intend to make you a better shooter, teach you sound tactical practices, and give you confidence with your weapon.

Stay in the Fight!!!!!!
Kyle Lamb

Why Chapter 5B and 15B?

The majority of this book was written while I was in Iraq. Over the last 5 years I tried to put together the book while also conducting combat operations. My normal schedule was to sleep during the day, conduct missions at night, and work on the book to wind down right before hitting the rack.

When I showed fellow operators what I was working on they were more than happy to give me their opinions. One of their opinions was to add a few more chapters. I added the chapters that they recommended. When it came time to get the book laid out, the odd chapter order immediately came up. When I told Matt, the fellow working on the book with us, he felt that it actually made sense to leave them as they were.

As you read these oddly numbered chapters, I hope you will take a minute to think about those that are serving abroad for our great country. I will never forget where I was and who I was serving with.

Kyle

CONTENTS

I. INTRODUCTION & MORE
   Warning. 2
   Dedication 3
   Acknowledgements 4
   Introduction 6
   Why Chapter 5b And 15b? 8

1. GETTING STARTED 13
   The AR Weapon System 13
   AR Background And History 14
   The AR Today 15

2. WEAPON SELECTION 17
   Barrels 17
   Barrel Construction. 17
   Stress Relieving 19
   Chrome Lining 19
   Choosing Barrel Twist 20
   Choosing Barrel Length 21
   Choosing Barrel Weight 23
   Choosing Barrel Fluting 24
   Trigger Systems 24
   Hand Guards. 25
   Accessory Rails 27
   Butt Stock Selection 27
   Chapter Summary 28
   On Your Own. 29

3. OH SAY CAN YOU SEE? 30
   Sighting Systems 30
   Sight Basics. 30
   Parallax 30
   Parallax Effects 31
   Detecting Parallax 31

   Focal Plane. 33
   Iron Sights 34
   Advantages Of Iron Sights 35
   Red Dot Sights. 35
   Eo Tech Holosight. 35
   Aimpoint 36
   Low Power Variable Scopes 38
   Trijicon Accupoint 38
   Schmidt And Bender 39
   Leupold CQT. 39
   Fixed Power Scopes 40
   ACOG 40
   Back Up Sights. 41
   BUIS (Back Up Iron Sights) 41
   Aperture Sizes. 41
   Windage Adjustment 41
   Elevation Adjustments 42
   Alternate Back Up Sights 42
   Jpoint 42
   Iron Sights 42
   How Often Do Scopes Fail? 42
   Mounting Your Scope 43
   Tricks Of The Trade 44
   Chose The Right Sights 44

4. HEADING TO THE RANGE. 45
   Loading /reloading. 45
   Speed Reloads 49
   Tactical Reloads. 51
   Unloading 52

5. MALFUNCTIONS 53
   Magazine Choices 55
   Ammunition Choices 56
   Malfunction Clearance 58

Immediate Action 58	Over Adjusting Or Chasing Zeros 95
Tap, Rack, Bang 59	Duty Ammunition 95
Remedial Action 62	Applying The Fundamentals 96
Analyzing Malfunctions 64	Positions 96
Double Feed 65	Analyze Your Target 97
Failure To Extract/stuck Case 66	Changing Zeros 97
Failures To Feed 71	8. BASIC BALLISTICS 98
Bolt Over Base 71	External Ballistics 98
Stove Pipe 72	Ballistic Terms 98
Additional Malfunctions 73	Why Study Ballistics? 99
Charging Handle Impingement 74	Bullet Rise 99
Bolt Override Malfunction 74	Trajectory 100
Dummy Rounds 76	Zeroing Your Rifle (Sighting In) 100
Transitioning To Sidearm 77	Zeroing Basics 101
5b. Tactical Slings 78	25 Meter Zero 101
Three Point Slings 78	Ballistic Chart M855 103
Single Point Slings 80	Ballistic Chart MK 262 104
Two Point Slings 81	Ballistic Chart Federal 62 105
Tightening The Vtac Sling 81	Ballistic Chart XM 193 106
Releasing The Vtac Sling 83	The Best Zero 107
6. FUNDAMENTALS OF MARKSMANSHIP . 87	Ballistic Software 107
Natural Point Of Aim 87	Wind Effects 108
Position 88	How Far Is Too Far? 108
Sight Alignment 89	9. STAND UP AND FIGHT LIKE A MAN! . . . 109
Trigger Control 89	Stance 109
Breathing 90	Hand Positioning 111
Follow Through 91	Forward Pistol Grips 113
Reading Your Sights 91	Mounting The Weapon 114
7. GETTING ON PAPER 93	Correct Head Positioning 114
Target Selection 93	Incorrect Head Positioning 115
Check Your Sights 93	Left Hand Safety Manipulation 115
Start Close 93	Muzzle Up Or Down? 117
Adjusting Sights 94	Weapons Retention 117
Shooting Groups 95	

10. WE DON'T LIVE IN A PRONE WORLD 120
    Natural Point Of Aim. 120
    Prone 122
    Body Positioning 122
    Weapon Positioning 123
    Foot Position. 123
    Support Hand 123
    Alternate Prone Positions. 124
    High Prone 124
    Rollover Prone. 124
    Elbow Positioning 125
    Support Hand During Rollover 126
    Reverse Rollover Prone 126
    Sbu Prone 127
    Rice Paddy Prone 130
    Sitting Positions. 131
    Conventional Sitting Position 131
    Spread Legged Sitting. 132
    Rocking Chair 133
    Stacking Your Feet 134
    Kneeling Positions 136
    Conventional Kneeling 136
    Seated Kneeling Position 137
    Stretch Kneeling 138
    Kneeling Around Corners. 138
    Standing Position Around A Corner . 141
    Junkyard Prone 143
    Rollover Kneeling. 144
    Broke Back Prone 144
    Tony Black Position. 145
    Using Barricades And Obstacles . . . 145
    Hand Guards Touching Barricades . . 146
    Barrels Touching Barricades 146

11. MEAT AND POTATOES 147
    Lead With Your Eyes 147
    Lead With Your Eyes 147
    Is Smooth Fast? 149
    Driving Your Weapon Drill 150

12. IS WEAK REALLY WEAK?. 151
    Eye Dominance 151
    Strong Side To Support Side 152
    Safety Manipulation 153
    Transferring To Support 154
    Support Side To Strong Side 157
    Keep Your Eye On The Target. 158
    Eye Dominance Tricks 158
    Practicing 158
    Tools For The Toolbox 160

13. FACING MOVEMENTS 161
    90 Degree Turns. 161
    RH Shooter Turning Left. 162
    LH Shooter Turning Right. 162
    180 Degree Turns 163
    RH Shooter Turning Left. 164
    LH Shooter Turning Left. 165
    RH Shooter Turning Right. 165
    RH Shooter Turning Left. 166
    Foot Movements 167

14. HE WON'T STAND STILL! 169
    Start Position 169
    Move Slowly 169
    Don't Slide 170
    Pause Stepping 171
    Moving Quickly 171

Read Your Sights 172	Light Discipline 195
Moving Obliquely 172	Visible Laser 196
Shooting While Moving L-R 172	Laser Rule Of Thumb 197
Shooting While Moving R-L 173	Parallel Zero 197
Zig-Zag Drill 175	Quick Zero Confirmation 198
Moving Around The Zig-Zag 176	NVG Shooting 199
Rear Security Technique 179	IR Scope Settings 199

15. TARGET DISCRIMINATION 181

15B. TRANSITIONING TO THE SIDEARM . 183
    Using The Safety 183
    Transitioning 184
    Transitioning To Sidearm 185
    Support Side To Sidearm 186
    Sling Adjustment 186
    Weapon Transition Practice 186
    Weapons Transition Drills 187

16. TERMINAL BALLISTICS 190
    Bonded Bullets 191
    Over Penetration 191

17. NIGHT SHOOTING 193
    What Works At Night? 193
    Light Selection 193
    Light Placement 194
    Using The Light 195

    IR Laser/spotlight 200
    Summary 200

18. WEAPON MAINTENANCE 201
    Basic Tool Kit 201
    Disassembly And Cleaning 204
    Cleaning The Upper Receiver 209
    Functions Check 214
    Springs 215

19. TRAINING AIDS 216
    Dummy Rounds 216
    Green/frangible Ammunition 216
    22 Conversion Kits 217
    9mm Upper Receivers 217
    Paint Marking Weapons 218
    Air Soft Weapons 218
    Training Guns - Blue/Red Guns . . . 219
    Visible Lasers For Training 219

1. GETTING STARTED

Let's start at the beginning. In this chapter I will introduce you to the AR weapon system, offer some guidelines for weapon selection and set the stage for the rest of this book.

If you have already bought or been issued an AR-type rifle or if you have considerable experience with this system, you may be tempted to skip this chapter. That is okay; you can always come back here for additional information if you need it. However, I urge you to at least skim the information in this chapter – no matter what weapon system you are using or how much experience you may have. I will make certain assumptions about your basic knowledge of terms and the function of an AR 15. It will be easier for both of us if we start on the same sheet of music.

The AR Weapon System

Let me start by establishing the tools of the trade that are the topic of discussion in this book. To get the most out of the instruction and training I'm offering, it is imperative that you possess or have access to an AR/M-16 type rifle or something similar. Although I will focus primarily on the AR system, any assault or combat rifle system will do. If you study and apply the techniques shown, I guarantee you will come away with knowledge that can be applied to achieve better marksmanship and better performance in those dangerous scenarios. You must be able to depend on the weapon you are carrying and your ability to operate it. In these situations, your target shoots back.

Unless you were embedded deeply in a foreign country (or you were living on another planet) during the 1990s and the early 21st century, you have most likely heard and read a lot of discussion about the so-called "assault rifle." For ten years certain weapons with a poorly defined look and capability were banned for sale to the general public. This silly ban did nothing to achieve lower crime rates as weapons of this type are rarely used for illegal purposes. This ban did, however, accomplish a couple of things. It cost manufacturers and suppliers a lot of money and it restricted the rights of American citizens to choose freely the type of weapon they own. I, for one, am glad those days are over.

The AR, at least in some of its variants, was included among weapons banned by the aforementioned law banning assault rifles. Designed as a combat rifle, the AR falls under the umbrella term "assault rifle" but a more common, and in my opinion more accurate name to call the AR is a tactical rifle. Tactical as per the handy dandy pocket dictionary I carry, means, skill of using available means to reach an end. The AR is a means to an end. In the end, we want the threat to be eliminated, we want no friendly forces to be injured or killed, and we want to be able to see our grandkids grow up. You should take this training very seriously and always strive to develop additional skills. You never know when these additional skills will be needed. I will continue to use the term tactical to describe the AR/M16 and other weapon systems I discuss throughout this book.

It is not so important which term we use to describe these weapons. What is important is the fact that with the assault weapons ban restrictions lifted you are able to purchase, build, or evolve the weapon that best fits your personal preferences or job requirements. In this section I will illustrate and define the various components of the AR system. Later in this chapter I will show you how to select these components for your own weapon.

AR Background and History

The AR weapon system is at once simple and complex. It is designed to work under adverse conditions. It is composed of interchangeable parts and can – with the right components – achieve extreme accuracy. How you build your system will be influenced by your application of the system: combat, home defense, competition, varmint hunting or plinking.

The AR, later to have its name changed to the M-16, was designed in 1957 by Eugene M. Stoner after being commissioned to do so by the U.S. Army. Mr. Stoner developed this weapons system relatively quickly when compared to the firearms industry today. It took him less than a year to design this rifle to the specifications he was given: weigh less than seven pounds and have the capability of automatic and semi-automatic fire.

The M-16 was quickly found to be superior to its predecessor, the M-14, in several areas. It was lighter than the M-14, could be controlled during full auto fire much easier, and allowed the soldier to carry a significantly larger amount of ammo within the same weight tolerances. The M-16 also worked well out to 500 yards when compared to its heavy weight challenger. Once the testing was complete, the United States Continental Command Board made a recommendation to adopt the M-16; however, it wasn't until 1962 that 1,000 rifles were sent to Vietnam for use and testing by U. S. Army Special Forces and Vietnamese soldiers. Figure 1.1 shows a typical 60s era M-16 of the type first deployed in Vietnam.

Figure 1.1: Typical Vietnam Era M-16, courtesy Fulton Armory

Initially the weapon was well received by the ground pounders. They felt they had finally received a piece of kit that enhanced their ability to fight in the jungles of Vietnam. This warm reception was echoed by Secretary of Defense

MacNamara, who intervened when Army Staff officers were reluctant to change from the conventional M-14. MacNamara decided to purchase 85,000 M-16s in 1963. They would later be sent over seas, to Vietnam. Meanwhile, stateside soldiers continued to use the M-14. You can see the difference in basic design from the M-14 rifle shown in Figure 1.2 M-14.

Figure 1.2: The M-14, America's Military Rifle Before the M-16, courtesy Fulton Armory

Everything appeared to be fine until 1966 when a sudden increase of unexplained jams occurred with the M-16. It has been said that many soldiers lost their lives due to these weapon stoppages. As it turned out, the jams had almost nothing to do with the weapon, but rather the powder that was substituted for the recommended specifications. It seems the extruded powder, IMR-4475 (Improved Military Rifle), had been replaced by a ball powder that created a much different firing sequence for the weapon, causing the bolt to move to the rear much quicker than intended. This additional bolt velocity resulted in sickening extraction problems. Some of these malfunctions were so severe that the soldier had to use a cleaning rod, from the muzzle, to remove the fired case.

The AR Today

Many things have changed since the Vietnam era. The M-16 has since evolved from its original design. The M-16A2, with its heavier, chrome-lined barrel, will stabilize much heavier bullets. This ability is due to the changing of the rate of rifling twist. Added to this version were brass deflectors to aid left-handed shooters as well as raised areas around the magazine release to prevent inadvertent magazine release. You will also notice that the forward assist has changed from a tear drop shape to a round shape. These significant but subtle changes aren't always obvious, as you can see from the photo in Figure 1.3 M-16A2.

Figure 1.3: The Newer M-16A2 Design, courtesy Fulton Armory

The latest AR craze is the M-4 size weapon. These battle rifles have a 14.5", relatively heavy barrel with a 1:7 twist, which means the bullet spins even faster than the M-16A1. They also have the collapsible stock to make a very compact, yet fine handling weapon system. Notice the tight-coupled design in the typical M-4 shown in Figure 1.4.

Figure 1.4: Typical M-4 Shortened AR Design, courtesy Fulton Armory

The most notable change is in the weapon's reliability. The magazines, extractors, springs, and overall parts quality have vastly improved. These improvements have made the AR one of the most sought after battle rifles ever.

To date over seven million AR type rifles have been produced worldwide. Not bad for a rifle that has spawned so much controversy.

2. WEAPON SELECTION

For most of us, selecting which new blaster to purchase can be a significantly emotional event, especially when you consider two very basic factors: how much does it cost and will my wife let me buy it? If your misses is kin to mine, she doesn't have a problem with purchases involving fine ballistic implements. You may not have a problem in this area either, but if you do, you've probably figured out how to hurdle this obstacle long ago. Either way, once you have the money and the wife's issues are quelled, it is time to build your ideal AR weapon system.

In this section I'll help you decide which specifications and accessories apply to your tactical rifle selection situation. If you are tweaking your design or attaching accessories simply for "cool guy points" you are on your own.

Throughout this book I'll assume you're using your AR for tactical employment. This approach should help you build a useable system that can be applied to most tactical situations across the ever changing tactical and battlefield spectrum.

The AR is a chameleon that can fit any of these applications. If you start with the right components at the top of the scale, the AR can always serve lesser functions. However, if you start at the bottom of the scale, asking your rifle for higher-level performance may not be realistic.

You can start designing your AR system however you choose, but for accuracy and performance, the barrel is a very important component. Let's start there.

BARRELS

There are a few basic barrel questions: How heavy, how long, which twist, and what type of construction?

I will discuss each of these issues in the following sections and offer some pros and cons for the various choices you'll find available. I'll start with barrel construction and twist and work back up the list.

Barrel Construction

When shooters talk about barrels, you just never know what tall tales you will hear. Shooters will regale you with tales of unbelievable shooting prowess, often never duplicated by the mere mortal man. Of course, during the life changing experience this shooter had, he will invariably place a large amount of praise on the components of that great day. I often wonder if it wasn't just "Lady Luck" stepping in to tighten their groups to this immortal size.

Some shooters prefer stainless barrels, others choose chrome-moly. Either material is a good choice. Both have their strong and weak points. Chrome-moly barrels are slightly cheaper to purchase in the beginning, and they are easier to machine. Stainless barrels, although more expensive, resist corrosion slightly better than chrome-moly. With either barrel you get what you pay for.

Whatever the material, barrels today are constructed with a variety of techniques; including hammer forging, button rifling, and single point cut rifling. Hammer forging starts with a short, fat barrel that is beat until it forms a barrel around a hardened mandrel. Hammer forging sounds medieval, but some top firearm manufactures still use this technique today; HK and Remington to name a few. See Figure 2.12

Figure 2.12 Hammer Forging photo, courtesy Precision Shooting

Button rifling sounds a little more sophisticated. With this technique the barrel blank is center bored and a hard, metal button is pulled through the barrel until the lands (raised areas) and grooves of the rifling appear. This technique is the newest and most widely used rifling technique today. Figure 2.1 shows the rifling button used to form the lands and grooves for an AR barrel.

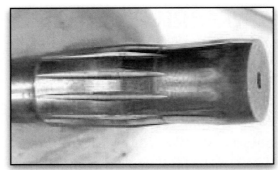

Figure 2.1: AR Rifling Button

Cut rifling or hook rifling involves the same center boring process as button rifling. After the center bore is complete, rifling is cut slowly into the bore with a single groove tool (This is sometimes called single point cutting). This process is very time consuming since several passes of the cutting tool through the barrel may be required for each groove produced. It also causes minimal stress to the barrel material and can result in very high quality and accurate work.

Broach rifling is a variation on the single point cutting method. A broach is a hard metal reamer constructed as a mirror image of the desired rifling. See Figure 2.2.

Figure 2.2: A Typical Rifling Broach

A number of successively larger cutting elements or rings are placed down the length of the broach so that all of the rifling can be produced with a single pass through the barrel blank. As the broach is spun into the barrel blank, each ring cuts a deeper and deeper groove into the barrel metal. After the last cutting ring passes through the barrel, the rifling is at the proper depth. Broach cutting speeds up the construction process, but broaches, constructed of very hard metal up to 16 inches long, are extremely expensive.

Most barrels for AR-type rifles are produced with button rifling, because it is cost effective and results in a barrel of reasonable quality and accuracy.

Stress Relieving

Stress relief of the barrel is another important process the barrel must endure before you get to apply your own sort of abuse. The barrel may be heated or cooled to excessive temperatures several times before it is properly conditioned or stress relieved. Cryo, a freezing technique, is the favorite of some. Others call this technique witchcraft. The stress relieving process is designed to allow the barrel to heat and cool with very insignificant shifts in point of impact as well as a return to zero when cooled. In layman's terms, stress relieving allows the molecules of the metal to relax; if they were not relaxed they would continue to pull one way or another to create slight imperfections in the trueness of the barrel.

The barrel building process sounds pretty technical, and it is. As simple as the concept sounds, building a quality rifle barrel is a complex and skilled process; but the bottom line is if you buy from a reputable manufacturer, no matter their technique, you will have a good product.

Chrome Lining

When you hear of chrome-lined barrels, shooters are referring to something different than chrome-moly or stainless. A chrome-lined barrel is a chrome-moly barrel that has had a chrome lining electrically adhered to the bore as well as the chamber. When chrome lining is applied you end up with a rough set of rifling, but remember what this weapon is for. If it is for tactical use, reliability may be more important than match grade accuracy.

The chrome lining on your mil spec barrel will degrade accuracy, but it will also increase reliability. The chrome lining on the chamber allows the extractor to easily pull the spent case from the chamber even when the rifle is extremely dirty. Of course, this is contingent on if you are using good brass and performing maintenance on extractors and springs.

If accuracy is your primary concern, you can find aftermarket barrels in chrome-moly and stainless steel. Either will get the accuracy job done; but, depending on whom you ask, one may last slightly longer than the other.

> **How Long Will a Barrel Last?**
> The amount of torture you decide to inflict on a weapon system will influence barrel life significantly. There is no exact answer. If the weapon is heated extremely hot during rapid fire training, or worse yet, full auto fire, the barrel will degrade significantly faster than if you were to allow cooling during your training. Even with chrome lined barrels, excessive heat will cause serious issues. If the barrel is heated to the extreme, small, soft spots behind the chrome lining can form, causing the chrome lining to degrade. The best advice I can give with regard to the life of your barrel is this: Train, train, and train some more. When you notice your accuracy falling away, have a gunsmith check your barrel, and, if needed, replace it. Barrels are cheap and life is precious; train hard so you can survive your next encounter. Don't get caught up in the superfluous worries about barrel life; worry about your partner's life instead.

Choosing Barrel Twist

What about twist? This depends on your preference in bullets and the application of the rifle. The term twist refers to how fast the bullet spins as it is accelerated down the barrel. Without rotation, the bullet would lose considerable accuracy. The rate of rotation coupled with the weight and type of bullet can influence accuracy.

Bullet rotation is described as how many times the bullet turns within a specified distance of travel. For example, a 1:9 twist denotes that the barrel's rifling makes one turn around the inside of the barrel in 9 inches. So the bullet spins once every 9 inches when fired. The significance for the shooter is how well the bullet will be stabilized.

A stable bullet will have enhanced flight characteristics. Enhanced flight characteristics translates into increased accuracy. A bullet that is not stable upon exit from the barrel may get so unstable by the time it hits the target that it may not perform as required to incapacitate the threat. It is also possible that the opposite situation may occur. An unstable bullet may tumble once it contacts the target, something that could increase lethality. The problem is that you just can't predict the final effect of a tumbling bullet on the target.

The military chose a 1:7 twist to help stabilize specialized bullets. This twist works well but it is not optimized for overall performance. If you are shooting a mil spec barrel, it may actually have a 1:7 twist, and if you have selected this particular barrel, accuracy is not paramount on your

AR checklist. I prefer to spec my barrels to shoot heavy 75 and 77 grain bullets extremely well, yet shoot light bullets, such as 55s, okay. To accomplish this, I use a 1:8 twist rate. The actual twist rate is 1:7.8, one turn in 7.8 inches.

A 1:9 twist is a great all around twist; however, for 75- and 77-grain bullets a 1:8 twist is noticeably better. Since the military has selected a 1:7 twist, other twist rates in mil spec barrels are hard to find. Mil spec is a reference used to indicate that this particular part will interface with any other mil spec part in your weapon system, a specification dictated by the military contract for that weapon system. This allows other manufactures to get involved in parts production if needed. They simply build the part to the contract specification.

You may have to compromise between mil spec and optimum twist for your particular application. As more heavy bullets become readily available, the shooting industry will begin to offer more twist choices for the tactical shooter.

Choosing Barrel Length

AR's with short little barrels – less than 14.5 inches long, say – are cute, but how effective are they? If your primary concern is compactness, short barrels are the ticket. On the other hand, short barrels significantly degrade velocity, which in turn changes the terminal performance of your bullets. Most short barrel ARs are not quite as reliable as their long-barreled cousins. (I mean reliable in the most basic tactical sense: will it go "bang" when you pull the trigger? That's a really basic requirement for a tactical rifle in my opinion). With the introduction of new gas-operating system designs from companies such as LWRC and HK, reliability with short barrels should improve. See figure 2.3-2.4A-B. These new gas systems allow the upper receiver as well as the bolt and bolt carrier section of the weapon to stay much cleaner because it does not allow any gas to be blown back into the action.

The LWRC system also allows you to retrofit your current system if desired. One benefit of the LWRC system is that the upper rail does not increase in height, therefore the transition to this system is seamless.

Figure 2.3 HK 416 Operating system. Courtesy HK USA.

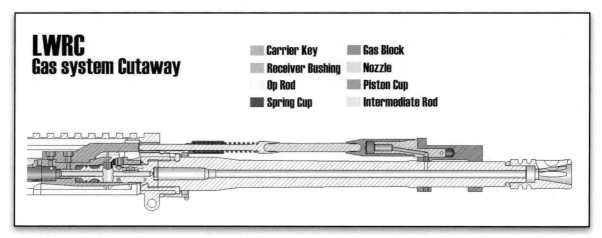

Figure 2.4A LWRC Gas system, cutaway. Courtesy LWRC.

Figure 2.4B LWRC Gas System. Courtesy LWRC.

During my travels I haven't seen a whole lot of real life situations in which a 10" barrel has a significant advantage over a 14.5" barrel. On the other hand, velocity is directly dependent on barrel length (among other factors), and I have seen several instances where velocity was a factor. For example, when you step from a confined space of a third world building into a long narrow alley that stretches for hundreds of meters, you probably count velocity and accuracy on the important side of your checklist. If a rifle is your primary weapon in confined quarters, such as a vehicle, a short barrel could be advantageous.

Normal barrel lengths are generally in the 14.5" to 16" range, although 20" AR systems also are available. If you work for a Government Agency, you probably will be issued a weapon in the 14.5" category. If you are a civilian or are required to purchase your own rifle for department use, it will generally be in the 16" or longer format. The reason being, due to current U.S. laws, 16" is the shortest barrel a private citizen can own legally without completing extensive paperwork and purchasing special permits.

As stated earlier, barrel length directly influences velocity, which influences in flight bullet performance, which ultimately influences terminal bullet performance. By terminal performance I mean the destruction the bullet causes to your target. If a bullet is designed to perform at 2700-3100 fps (feet per second) and your barrel length launches the bullet at 2500 fps, Ray Charles can see what is going to happen here: under penetration due to lack of velocity, or quite possibly over penetration because the bullet doesn't expand as designed. We will discuss Terminal Performance of your bullet in Chapter 16: Terminal Ballistics.

When you decide on barrel length, ensure the system is reliable with your ammo and that the preferred bullet design will still perform as you wish at the velocity you can achieve with your chosen design.

Choosing Barrel Weight

Now that you have decided how long or short your barrel needs to be, you must decide how heavy you want the barrel. Most AR barrels are too heavy. Realistically, if you plan to use your rifle in a tactical situation, how heavy does the rifle need to be? In the real infantrymen's world, lighter is always better. This, of course, depends on whether or not the barrel weight or contour can handle the punishment you plan to inflict. If the weapon is for close quarter engagements, can it withstand the heat and abuse of prolonged, rapid-fire training? True, few of us will ever become decisively engaged, but if you do, the encounter would lead to significant barrel heat.

So what is it going to be, a light barrel for combat or a slightly heavier barrel to withstand the rigors of tactical training? If most of your usage will be training with the slight chance of entering a combat situation, a more rugged barrel would be the right choice. If you know that combat will occupy most of your rifle toting time, it probably is okay to sacrifice a little durability in favor of lighter weight.

There are several custom barrels on the market that really feel great for carrying around all day. They also have accuracy standards that far exceed mil spec equipment.

One such barrel is the JP Enterprises, Lightweight Barrel (www.jprifles.com). This barrel is significantly lighter than others on the market, without sacrificing accuracy. JP is able to achieve this seemingly contradictory feat by building a precision, yet lightweight, barrel, and insuring there are no outside influences on the barrel. This is accomplished by using free floating barrel systems on all of their rifles.

JP also has designed a heat sink that attaches to the barrel and helps to disperse heat quickly in a situation that requires a high volume of fire. This is a finned device, much like you would see on an air-cooled motorcycle engine. It is constructed of aluminum, which is a great heat conductor and extremely light. The fins create a larger surface area, allowing for much quicker heat dissipation. Figure 2.5 shows a JP rifle with the lightweight barrel and heat sink.

Figure 2.5 JP Rifle's Lightweight Barrel and Heat Sink

The addition of the heat sink to any light weight barrel will help to reduce barrel wear by lowering temperatures. It will also help to increase accuracy during prolonged shooting situations. Excess heat causes increased shot dispersion. Without the heat, shot dispersion is significantly reduced.

Pick the weight that fits your needs. You may need to have several upper receivers for changing situations. Find one that works for you. I am a fan of lighter barrels and lighter rifles; anything lighter is better. That is, if it will perform to the standards you require for your job performance.

Choosing Barrel Fluting

Next to enter the equation is the concept of fluting. A fluted barrel has had a series of symmetrical, longitudinal cuts made in the exterior of the barrel. See Figure 2.6.

Figure 2.6: A Fluted Barrel Design, courtesy Mike Haugen

Does it help? There are certain times fluting can provide positive results. If you are trying to build a long barrel rifle that is relatively light, fluting may be the answer; however, fluting is definitely not a necessity. When dealing with extremely accurate barrels, fluting can sometimes cause irregular performance from barrel to barrel. As the barrels heat, they tend to shift erratically due to the flute design.

Some manufacturers have started to be more innovative with barrel lightening. Instead of flutes they are using dimples. These dimples help to reduce weight while maintaining good rigidity. Does it work? Only time will tell. We need more field experience to know for sure, but it does sound like a good idea.

Trigger Systems

Going hand in hand with barrel selection is your choice of trigger. Several manufacturers have modified the AR trigger to help you achieve better accuracy and improve follow through. These match triggers may be adjustable for take up (the amount of movement in the trigger before you release the sear and the hammer falls), over travel (the amount of movement of the trigger after the hammer falls), and of course weight (the amount of exertion applied before the trigger/sear releases the hammer).

Several of the available triggers are two stage designs. A two-stage trigger requires significant movement before you reach the actual trigger pull, or movement of the trigger, sear, and hammer. This will help long-range accuracy, but up close, a two-stage trigger is definitely slower.

A nifty new addition to the AR trigger market is drop-in trigger modules, available from McCormick and JP Enterprises. See Figure 2.7.

Figure 2.7: JP Drop-In Trigger Module

These systems let you quickly drop a pre-adjusted module into the lower receiver, replace the trigger and hammer pins, conduct a functions check, and be off and running. Both of these triggers are single stage triggers. The JP, however, will allow you to adjust the safety for proper function. Once you have installed an after market trigger, you must ensure that the safety cannot be engaged with the hammer forward; if it will, you need to immediately get this fixed or install a different trigger system. If it is possible to place the weapon on safe with the hammer forward, this could have severe effects in a tactical situation. If you were to have a malfunction that allowed the hammer to go forward – and you were able to place the weapon on safe – this would not allow you to pull the bolt carrier to the rear to clear the malfunction. With a proper functioning AR weapon system, the safety should not move into the safe position with the hammer forward.

If you decide to use a match trigger that is not modular, have it installed by a competent AR gunsmith. Ensure all adjustment screws are secured with Loctite® or another adhesive product, and once the trigger is adjusted, do not soak these parts in solvent. This may cause the adhesive to slip during repeated firing thus causing a trigger malfunction.

How light should this tactical trigger be adjusted? I recommend at least 4 pounds. This will help you to feel the trigger with gloves on as well as prevent an accidental discharge in a stressful situation. In addition, your rifle's trigger weight should be as close to your pistols pull weight as possible.

Hand Guards

Barrels and triggers have a direct effect on the accuracy of your weapon system. The correct hand guard system is another item tactical shooters frequently overlook. The hand guard, or fore stock, is the piece on the front of the rifle that surrounds the barrel and provides a grip for your forward hand. It also provides protection for the gas tube as it runs along the top of the barrel.

A free float hand guard, a tube that does not touch the barrel like regular AR hand guards, will significantly enhance your rifle in more ways than one when used for tactical applications. First, accuracy will be enhanced because there is no outside interference placed on your barrel. The barrel is allowed to move freely as you fire. Outside influences such as concrete building crowns, car hoods, and metal light poles, when used for support, can cause significant zero shifts. These shifts are not significant at 25 yards, but may cause you to miss head shots at 100 yards. With a non-free floated barrel system, using a sling to apply pressure to the barrel, I have been able to move the bullets impact as far as 4 inches at 100 yards. This is significant when you consider the size of the vital organs we are trying to hit with this weapon. Being able to use your sling as a shooting aid, with the free float tube, also gives you an increased advantage when shooting increased distances.

Your hand guard should also attach securely to the barrel nut; there should not be any slop when attached. If you decide to attach IR lasers and lights, it is crucial that your hand guard does not move after your rifle's accessories are zeroed. You must also take into consideration how quickly the hand guard transfers heat to your support hand during prolonged shooting.

There are several great systems on the market such as the Noveske Rifleworks free float tube, shown in Figure 2.8, and there are several that are way over priced for the quality you receive. John Noveske not only builds quality tubes, but builds top of the line carbines. I helped JP Enterprises design the Viking Tactics Tube, shown in Figure 2.9, due to the shortcomings of other systems I have used in the past.

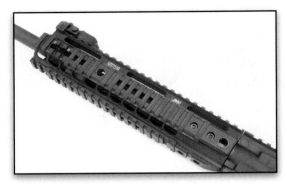

Figure 2.8: Noveske Rifleworks Hand Guard

Figure 2.9: Viking Tactics Hand Guard

Notice that the Viking Tactics design provides a light weight, well ventilated design, with the ability to remove any rail sections that are not being used. This system also maintains your zero for any item mounted on the hand guard.

If you decide to use a non-floated hand guard system, don't blame the rifle when accuracy isn't to the highest level or you have unexplained zero shifts from different positions.

Accessory Rails

Lastly, are there enough accessory rails to attach what you need, or are there too many? Accessory rails allow you to mount BUIS (back up iron sights), IR lasers, flashlights, bipods, and forward pistol grips.

There are two types of accessory rail systems in common use with the AR weapon system: Picatinny rails and Weaver rails. Both rail systems are designed to perform the same job, and, in general, accessories designed for one system will fit the other. However, this is not always true. And, while you can find quality products for either system, the military standard is the 1913 Picatinny system and for the purposes of this book I'll stick with Picatinny whenever I talk about accessory rails.

I have found that if the rails are removable, the tube will be easier to carry and you can remove the cheese grater effect, the tendency for the rough edges of the rail to rub your hands raw during extended carry situations. Some are so sharp that they can actually wear through your gloves. It is nice to have the ability to mount the rails at the 3,6,9, and 12 o'clock positions; however, you should be able to mount rails at positions between the 3,6,9, and 12 o'clock positions for increased flexibility of the system. If the rails are not removable, the tube will feel too large as well as uncomfortable during extended carry conditions.

> ### Accessory Rails
>
> *The Picatinny rail system is the result of a military specification (MIL-STD-1913 (AR)) adopted on February 3, 1995. This published standard set the dimensions and tolerances for any accessory mounting system that would be adopted by the military. The military rail system takes its name from the location where this system originated, the Picatinny Arsenal in New Jersey.*
>
> *The Weaver system is similar to Picatinny, but it has smaller grooves and inconsistent spacing between grooves. If you install Picatinny rails on your AR weapon, you will normally be able to mount accessories designed for the Weaver system, but the reverse may not be true. Weaver rails may not allow you to successfully mount accessories designed for the Picatinny system. For best parts modularity, choose Picatinny rails.*

Butt Stock Selection

There are no hard and fast rules when selecting the right butt stock for your tactical carbine. If I could have any butt stock on my rifle at all times, it would be the standard M16A2 butt stock. This stock allows me the most stable platform to engage targets. When we consider the tactical side of things, my selection changes, and now there are other considerations. If I am going to be working in an environment that supports a more compact weapon system, a collapsible stock will be more important. See Figure 2.10 Vltor Stocks.

Figure 2.10: Vltor Buttstock, courtesy VLTOR

Several of the newer stocks on the market are very tricky, but what do they really offer. I do like the fact that several manufacturers have adapted battery compartments to their stocks. This is a great addition to the shooter's tool kit. Other stocks also add additional length of pull, or shortened length of pull. Either way, you just need to find a stock that fits.

Normally shooters start with their stocks collapsed at least slightly. After a day or two of training, they normally will start using a longer stock. Some stocks are permanently shortened; not having the ability to lengthen your stock is a huge disadvantage. These short stocks do not allow most shooters to completely control their rifles; they cannot get a good stock to cheek weld and are not able to get the balance or center of gravity where it is needed. The short stock will make the gun feel unmanageable and take away much needed power to drive the weapon from target to target. A fully collapsed stock also makes your face hit the charging handle, or it can cause the scope to come back and hit your face.

The stock you choose should also be able to quickly collapse in order to help clear malfunctions. If you have a stuck case, for example, you will collapse the butt stock before you hit the butt of the weapon on the ground. If you do not collapse the rifle, you could break your buffer tube, making the rifle a club.

To check for proper fit of my butt stock, I first mount the weapon to make sure my nose isn't hitting the back of my scope. If it is, then I will extend the stock. If I feel like I really have to stretch to get my nose to touch the charging handle, then I will slightly collapse the stock.

Chapter Summary

This chapter is the beginning basics of gaining or improving tactical rifle skills and expertise. I've given you a little background on the AR weapon system, discussed the basics of AR system design, and shown you how to evaluate system components so you can build a weapon of your own to match your professional or personal needs. In the next chapter, I will discuss the beginnings of achieving basic accuracy, including how to hold the rifle, using the sights, and sight alignment.

On Your Own

If you already own or have been assigned a tactical weapon system, now is a good time to evaluate your system in light of the discussion in this chapter. Perhaps you haven't given much thought to what type of barrel you have or how the hand guard is designed. Take inventory of your current weapon. Understand what you really have and make some notes about your weapon. What is the rifling (twist)? Is yours a heavy barrel or a light weight? Do you think it is optimized for accuracy or for ease of carry? What is the barrel length and what kind of sights does it have? Is the barrel Chrome lined?

By understanding precisely what you have to work with, you will be better equipped to get the most out of the accuracy and weapon handling instruction that I will cover in the rest of this book.

If you don't already have an AR weapon system, use this chapter as a guideline for acquiring such a system.

Later in this book I'll give you specific exercises and techniques to practice. Repeating tactical skills will help you to learn these skills and develop muscle memory, making them second nature. Repetition will also help you to internalize these skill sets.

For now, inventorying the components of your weapon system will help you internalize details about what you have and how you will employ this weapon system.

3. OH SAY CAN YOU SEE?

SIGHTING SYSTEMS FOR THE TACTICAL CARBINE

I've talked about the foundational components you can choose to make your AR system fit your operational needs as well as your budget. One important component we didn't discuss was your sighting system. The original military M-16 was equipped with open iron sights. For daily carry in a battlefield environment, basics are better. The AR iron sight system is rugged, quick to acquire and put on target, requires no maintenance, adds very little weight to the rifle, and is easy to learn to use. The M16A2 offers the shooter the ability to adjust his sights quickly for long range engagements by using the rear elevation wheel.

Iron sights are functional in most lighting conditions and are extremely accurate to at least 200 yards for most users, but modern technology affords a number of sight enhancements for specialized operational conditions. Specialized sighting systems can replace the basic iron sights or work in conjunction with them.

In this chapter I'll introduce you to available tactical sighting systems and show you how to evaluate them for your own applications.

SIGHT BASICS

I'll talk more about practical sight use in Chapters 6 and 7. For now, I want to show you the basics so you understand how various sight systems work and so you have a foundation for choosing sights for your particular tactical needs.

Parallax

Parallax is often misunderstood in the shooting world. Parallax is a condition that exists when the image being viewed through a rifle scope doesn't fall squarely on the reticle. Most big-game hunting scopes are set parallax-free at 100 yards, and parallax present at longer and shorter distances cause such a slight sighting error that it's of no concern to the shooter. However, on target and varmint scopes that are used for shooting very small targets, often at long range, parallax can be the difference between a hit and a miss. Varmint and target scopes should be equipped with an adjustable objective lens so that parallax can be removed at various distances. By moving the objective lens axially, the image is focused on the reticle plane and sighting error is eliminated. Tactical scopes have a similar adjustment that is normally located on the left side of a scope. See Figure 3.1. It is believed by some that this is to focus your reticle; however, it is actually there to delete any parallax in your scope.

FIGURE 3.1 Leupold with Objective Focus, courtesy Mike Haugen

How does Parallax affect the Tactical shooter?

Some scopes on the market are made specifically for shotguns and, by design, the parallax free setting is at a relatively close range, normally 50 yards or less. This can cause significant parallax when the tactical shooter tries to not only zero their rifle, but also to engage small targets; for example, head shots at ranges beyond 100 yards. Another problem is the adaptation of non- magnifying, red dot, pistol scopes to your rifle system. Several of these scopes experience parallax issues when fired at ranges beyond 25 yards. I have experimented with several red dot scopes and they were almost worthless when engaging targets at 100 yards and farther. As red dot technology advances, this becomes less of an issue for the tactical shooter. It is easy to test your scope to ensure you don't have parallax problems.

Detecting Parallax

Parallax can be detected by moving your eye left and right, up and down, as you look at the target through the scope. You should have your rifle secured on a shooting bench or with sandbags to eliminate any movement of the rifle during this test. If the image seems to move in relation to the reticle, you have parallax. A fuzzy, out-of-focus image does not indicate parallax, but simply points to improper focusing of the eyepiece for the user's eye. Some shooters find this insignificant; however, if you are shooting 300 yards and you have 5 inches of parallax in your scope, ammunition that will shoot 5 inches at 300 yards, and a rifle that shoots 3 inch groups at 300 yards, this could equate to 13 inch groups. If you eliminate the poor ammunition, you would still have an 8 inch group. Eight inches doesn't sound like a very big group, but consider the consequences when trying to make a difficult head shot.

Eliminating or at least dealing with PARALLAX. We can't eliminate parallax, but we do have certain ways to better deal with these effects. If you are shooting a red dot scope that does not magnify, you are in luck. To help eliminate this parallax we need to have a consistent, stock to cheek weld. If you use one of the obscure shooting positions we discuss in Chapter 10, this may be difficult if not impossible.

To remedy this situation we need to align our dot with our front sight post. If you have a flip up front sight, then you will see a difference in your group sizes if you use it as a reference during the zeroing process. This is also another reason for having the dot align with the iron sights and not sit above. For a tactical carbine, BUIS (back up iron sights) are a must. Now you have more than dead batteries or a broken scope, as a reason for BUIS's.

MOA and how they apply to tactical shooting. *Minutes of angle (MOA) are commonly used in the shooting world as a point of reference. A minute of angle is 1.047 inches at 100 yards. For normal shooter application we use a simplified figure of 1 inch at 100 yards. To keep this in perspective, the difference between using the exact 1.047 and 1 is 0.47 inches at 1,000 yards. This is not significant when using a tactical carbine. So, if we say a reticle has a 1 MOA dot, this dot would cover 1"inch at 100 yards. This is also relative to the click adjustments on your scope or iron sights. A sight with 1/4 MOA clicks would require 4 clicks to move the bullet 1MOA or 1 inch at 100 yards. This same adjustment would move the strike of the bullet 1 MOA at 200 yards which is 2 inches. If you have ammunition that shoots 3 MOA, that would translate to 3"inches at 100 yards, 6"inches at 200 yards, 9" inches at 300 yards, etc. MOA adjustments can become confusing when trying to attain a zero at ranges closer than 100 yards. For example, if your optic has 1/2 MOA adjustments and you are shooting from 25 yards, you want to move the bullet strike 1 inch. This would require 8 clicks of adjustment. An easy way to remember this conversion is by using the 100 yard adjustment and doubling the number of clicks for 50 yards, and doubling again for 25 yards. So, with a 1/2 MOA scope to move 1" at 100 it will require 2 clicks, 1" movement at 50 Yards, 4 clicks, and 1" at 25, 8 clicks.*

Mildots, what are they and how do they work? *The Mil-Dot reticle was designed around the unit of measurement called the milliradian or mil. The dots of the mil-dot reticle were designed to allow the shooter to estimate range to a target that is of a known size, hold over targets with the mil-dots as a reference, and gave a recognizable lead for moving targets. There are differences in what exactly the mil-dot is equal to. Some* folks in the Army say each mil is 3.375 MOA (minutes of angle), whereas the Marines use 3.438 MOA for each mil. Either way if you use 3.5 MOA per mil, you will be close enough for government work. Even the scopes tolerances at the manufacturer are farther off than the Army and Marine differences. Using 3.5 will at least make your math slightly easier.

When using the Mil-Dot reticle for ranging, you simply place the reticle over the target and align one end of the target to the flat of the reticle, and move along the reticle line, counting the number of mils. You will need to make a guess as to where the target ends if it is between mils. The more accurate your mil reading, the more accurate your range estimation, and, most importantly, the more accurate your shooting.

If you need more information on the Mil-Dot formulas for conversion there are several web sites and companies that will supply you with this information. See Figure 3.2 MILDOTS

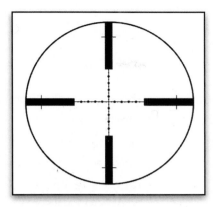

Figure 3.2 Mildots

Focal Plane

In the old days, or at least 10 years ago, no one in the US mentioned focal plane. This has changed in recent years. To simplify matters the reference to first or second focal plane, as it pertains to the application of ballistic tools, indicates what relation the reticle will have as magnification or power is increased or decreased.

Second focal plane scopes, normally any inexpensive, variable powered scope, do not allow the reticle to stay in perspective as power is changed. An example would be the use of a mil-dot scope. If you look at a 39" E-type silhouette at 200 meters, you should have 5 mils as a measurement from top to bottom on the target. If you change the power of the second focal plane scope, the mil reading will change, thus making a range estimate inaccurate with this scope if it is not set on a specific power.

On the other hand, if you employ a first focal plane scope, it does not matter which power you choose. The mil-dots will stay in perspective. This will allow for more accurate range estimation.

One drawback with the first focal plane scope is the inability to see your reticle clearly at low power settings. This becomes increasingly important as you get closer to your target. At 100 Meters, it may not matter; at 15 feet, it could slow you down severely if your retice is not easily identifiable. Several scopes on the market today have addressed this issue by adding a thick ring around the center reticle. This ring appears extremely large at the high power setting; however, when you dial it down for up close and personal shooting it will help to have the thick ring, which will now be much smaller, available for quick target acquisition.

If your AR is going to be employed as a sniper/varmint scope, score yourself a high quality, first focal plane scope. If you decide seeing the reticle quickly is more important, set your sights on the second focal plane scope. Either way, with practice you can effectively employ any high quality gear, regardless of the reticle and focal plane issues.

The addition of illuminated reticles will also increase your speed at close range, no matter the focal plane. This applies only to those scopes that allow the shooter to brighten the dot sufficiently for sunny days, as well as when a weapon light is used to illuminate a target at night.

IRON SIGHTS

Iron Sights as Primary sighting system

Iron Sights have become more and more rare as a primary sighting system in the tactical community. This is largely due to the reliability of optical sighting systems, as well as increased battery life for red dot sights.

If you decide to use iron sights for your primary system, here are a few things to think about. Can you see your sights at night? If not, you may need to add a Trijicon Tritium front sight.

Figure 3.3 Picture of Trijicon Tritium Front Sight

Is your sighting system adaptable for your tactical use, ie, does the sight allow you to zero at the range you need. An example of this would be the M4 weapon system. Initially this system did not have a long enough front sight to allow the rifle to be zeroed at 100 yards. This doesn't sound like a big deal unless you shoot a 100 yard zero, then it can be very disturbing. This issue has since been fixed and you should not experience this with a new system.

Another issue with selection of your Iron sighting system is the ability to switch from your small aperture to the large without a large deviation in zero. Issue sighting systems on the M4, as well as M16A2, have two apertures. One is intended to be used for close range (0-200 Meters), as well as zero. The other is intended for long range. I personally prefer to have both sights at the same height. By this, I mean the center of the aperture is in the same location with both sights. If you have a slightly off center hole, as you will with the A2, in one of the apertures it can create more confusion than needed. It will also make it more difficult to attain zeros when changing between apertures. If you decide to use these sights, always remember that the close range sight sits slightly lower than the long range. This equates to lower impact at decreased ranges. See Figure 3.4.

Figure 3.4 Large and small aperture beside each other

Advantages of Iron Sights

There are several advantages iron sights offer that cannot be overlooked. First, you do not need batteries for iron sights. Also, they are very durable, as opposed to some magnifying scope systems. If you do not plan to engage targets at extended distances, iron sights may do you well. Irons also work well for soldiers who may conduct maritime operations. I trust very few scopes when they have been exposed to salt water.

RED DOT SIGHTS

The latest innovation in the optical arena has to be the Red Dot sighting system. These scopes have been in use for some time but manufacturers continue to improve their products. Battery life is extended to almost unlimited, or once a year, which I find better than every couple hours of use. Parallax has almost been eliminated and durability is above and beyond what you would expect from a sighting system. Several manufacturers have improved the intensity of the sight so as to make it more visible in bright sunlight. They have also decreased the dot size for more accuracy potential at extended distances.

EO Tech Holosight

EO Tech has led the way into the 21st Century with the design of the Holosight. This system is a holographic projection design, similar to the heads up display in a fighter jet. Your aiming point, or dot, is projected onto a vertical screen. Several advances have been made since the introduction of this system. IR (Infra Red), the capability to lower the intensity for use with Night Vision devices, has given soldiers and law officers the option to use the sight more effectively at night.

The Model 552.A65 has also been modified to accept AA batteries. This modification makes battery acquisition much easier than trying to find N cell batteries. AA batteries also significantly increase battery life.

The newest version of the EO Tech, the Model 553, will accept CR123 Lithium batteries, the same batteries used in most quality weapons mounted lights. This eliminates the need to carry so many different types of batteries stashed somewhere on your rifle. The Model 553 also comes with an integrated throw lever mount, this mount is not the best mount available but it does allow the shooter to quickly mount or remove the device. I do not see a need to quickly take this sight off unless you have another optic that can be put in its place for a specific mission.

With the EO Tech, you also have the choice of several reticles. Normally, these sights come with the A65 reticle, which is a 1 MOA (minute of angle) dot with a 65 MOA circle around the dot. This reticle allows the tactical shooter to quickly acquire the outer circle at close range, as well as provides a well defined aiming point when more accuracy is needed.

It has been my experience that this sight is the fastest sight available for close range target engagements. This is largely due to the fact that you are not looking through a tube, as with most other red dot sights.

The EO Tech Holosight has also proven to be a very rugged design that holds zero, even when abused beyond normal bumps and bruises applied to issue gear. These scopes will even work when most of the glass in the upright display is gone. As long as there is enough glass to allow you to see the dot, it will work. See Figure 3.5.

Figure 3.5 Holosight with broken lens, reticle still visible, courtesy EO Tech.

Aimpoint

The Aimpoint was the first non-magnification, red dot scope on the market really worth owning, but the first Aimpoints left a lot to be desired. The Aimpoint has since evolved into a very rugged, bright, reliable, red dot scope. The battery life of the Aimpoint has reached into the thousands of hours, so there are not many worries there. The Aimpoint also offers low light, night vision settings, to help increase the effectiveness of night fighters. The Aimpoint gives the shooter several mounting options since it is a regular tube scope that does not have an integrated mounting system. Recently, Aimpoint introduced a 2 MOA dot scope. This has been long awaited in the tactical community. Until now, using the Aimpoint for extended ranges was impractical due to the large dot. Trying to superimpose this large dot on a 300 yard target head was a frustrating experience, to say the least.

Aimpoint has also introduced a 3 power extender (See figure 3.6) that can be placed behind the Aimpoint as well as the EO Tech Holosight. This extender does not change the zero of your scope; it simply gives you the extra magnification that is sometimes needed. The 3 power also has a unique mounting system that allows you to simply twist the mount to remove your adapter from behind the scope. The only drawback to using this adapter with an EO Tech is the fact that you need to be able to access the rear of the EO Tech to turn it on and off, as well as for intensity adjustments. With the adapter mounted close, this is a bit tedious, as well as the fact that you would have to hang onto the scope once it is removed. To remedy this problem, Samson came up with the flip mount for this extender. (Figure 3.8) This flip mount allows the shooter to quickly flip the sight out of the way with the press of a button. When you are ready to use it again, simply flip it back into place.

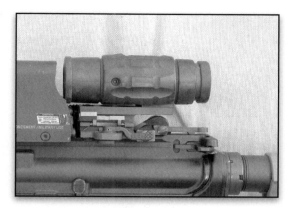

Figure 3.6 Aimpoint 3x extender

Figure 3.7 EO Tech extender

Figure 3.8 Samson flip mount

Figure 3.9 Larue Tactical Aimpoint 3x Mount

Battery Storage: *With the addition of battery powered lights, lasers, and sights for the tactical carbine, it is necessary to have a means of stowing batteries for easy access on your firearm, should your batteries die. Several devices have been made available to aid you in easy storage and access. When storing these batteries, it is also a good idea to waterproof the batteries to keep them fresh for emergency use. Another technique is to simply tape two batteries to the top of your sight. With two N-cell batteries taped to the top you will have immediate access to fresh batteries. Always remember to change your spare batteries at the same time you replace the batteries in your sight.*

LOW POWER VARIABLE SCOPES

Trijicon Accupoint

The Trijicon Accupoint (Figure 3.11) is one of the most overlooked scopes on the market. Not only are these scopes made by a great manufacturer, but they work extremely well. An additional benny is the fact that they are not that expensive, by today's standards anyway. This scope has a reticle that is illuminated during the day by a fiber optic system that draws light down into the reticle without the use of batteries. During night operations, the reticle is illuminated with tritium. This scope also has a shield that can be rotated to turn off the day time reticle for increased accuracy at extended distances. The only slight downfalls I have seen with this scope are the 1" tube diameter, which doesn't seem to cause any problems, I just like 30mm tubes, and the fact that it won't quite go down to a true one power, it only will go as low as 1.25 power. This also doesn't seem to cause any issues up close; I credit the bright aiming point. The triangle is extremely bright and eye catching.

When the Accupoint is used at night, it will make you a believer. This scope is great for close target engagements as well as shines when engaging at extended distances. Light or no light it seems to work well.

The Accupoint also has extremely long eye relief, a flaw Trijicon should remedy.

Figure 3.13 Trijicon Accupoint

Schmidt and Bender 1.1-4X Short Dot

The Schmidt and Bender Short Dot scope is one of the better quality pieces of tactical glass on the market. This scope was designed to fill a need that red dot scopes couldn't, the ability to not only target discriminate at extended distances, but also be more precise in a long distance, tactical situation. The Short Dot also needed to have a red dot for use in CQB distance engagements. There are several good things about this scope and a few that aren't so good.

First and foremost, the quality of the optic is second to none. This will allow you to see much better definition at the extended ranges where this scope shines. The click adjustments are very crisp and have always been accurate with the Short Dots I have used. This scope also has a 30mm tube which I favor over the one inch tubes. The shortcomings of this scope are really not the fault of Schmidt and Bender. The original Short Dot did not have locking turrets on the adjustment knobs; this was an oversight that has caused the Short Dot to now come with extremely large locking knobs. The original scope should have had a normal turret cover and smaller knobs. Oh well, maybe their next model will be better. The scope is also very heavy for an AR type rifle; the extra weight is worth it if you are in a open desert environment though.

The red dot that can be activated in the Short Dot is extremely bright; this is especially eye catching when trying to shoot fast at room engagement distances. The dot is a bit large at 5.5 MOA but should not affect your close range shooting.

Figure 3.14 Schmidt and Bender Short Dot, courtesy 10-8

Leupold CQT

The Leupold CQT is also a unique scope. This scope has blazed the path for other companies to follow. The CQT offers a decent reticle that can also be illuminated via battery power. The reticle in the CQT is not conducive to long range accuracy with the huge center dot, but up close and personal it is fine. The illuminated reticle is also not really illuminated unless you are in an overcast or night time environment. I am sure these shortcomings will be fixed eventually when there is enough call for the changes. This scope will actually dial down almost to a true 1 power, this helps to enhance your close range shooting, at least when speed is involved. The CQT scope is also very rugged; I have seen these scopes take abuse that the Samsonite Gorilla would be proud of.

Figure 3.15 Leupold CQT

FIXED POWER SCOPES

ACOG

If you are friends with UBL (Usama Bin Laden), you may not know of the enormous popularity the ACOG has developed. Most cave dwellers don't.

The ACOG is the most popular fixed power combat scope ever. There are very few reasons not to use this scope when hunting dangerous, human, game. The ACOG is produced by Trijicon and has a great reputation for its durability. I have seen a few ACOG reticles break but to caveat this comment I will say this: the reticles that broke were on earlier model scopes, and, to make matters worse, they were being employed on machine guns. Not 5.56 MG, but 7.62 and higher.

As always, when soldiers break a Trijicon product, they don't just fix the disabled item. They re-engineer so as to provide the war fighter an improved piece of kit.

The ACOG also allows tactical shooters to perform tasks that iron and red dot scope shooters cannot. One such task is being able to look back into the shadows of a house from extended distances to identify very bad things in wait. This scope can also be used to perform this function during hours of darkness. It may not sound significant until you hear bullets crack by your head and you have no idea where they originated; it is at this time that it crawls quickly to the top of your list.

The NSN or military version of the ACOG comes equipped with iron sights built into the top of the scope. This would fit into the, "not needed", category. It could also fit onto the, "don't work" list. I am sure someone thought it was a nifty touch for a tactical scope, but they are off base a bit. I have used a JPoint sight attached to the top of this particular scope, now that is a good idea. I don't like the fact that you have to change your stock to cheek weld to see through the JPoint, however it is relatively quick at close range. Significantly faster than looking through the fixed, 4 power scope. See Figure 3.16.

Figure 3.16 ACOG with JPoint attached

When you start spending more and more time with an ACOG, it will definitely grow on you. The reticle is designed to accommodate the ballistics of military issue ammunition, helping to take a lot of guess work out of long range target engagements.

BACK UP SIGHTS

BUIS (Back Up Iron Sights)

Back up iron sights are a phenomenon resulting from the influx of optics into the tactical market. Before red dot optics were introduced, you very rarely would see a back up sighting system. This even includes Sniper rifles. It amazes me when you ask tactical shooters why they have back up sights, normally they just say, "because, you must have back up sights". Do you? I am also a big fan of a secondary sighting system, however, when was the last time I actually had to employ them. Never. In the last few years, battery life has been extended a great deal; to the point that the excuse of batteries going dead just doesn't hold water. On the other hand, BUIS are so small and light weight, why not?

Aperture Sizes

An important issue to remember when purchasing BUIS is that they are for Back Up, or Emergency. Most BUIS's on the market are highly over engineered. The first over engineering is the addition of multiple apertures. This sight is for back up and you should select a rear sight that offers a good all around aperture size. If the situation dictates that your emergency sight is employed, it should work in the worst case scenario; which I believe would be an imminent threat, 100 yards and closer. Numerous sights have very large apertures or extremely small apertures. I would select a large aperture over a small aperture. If need be, it can be used as a ghost ring sight. If the sight has several or at least two apertures similar to the A2 sights, you will need to know the difference in zero from the large to the small aperture. These sights are designed for the large aperture to be used from 0-200 Meters and the other, smaller aperture, to be used for extended ranges.

Windage Adjustment

Several back up rear sights also offer A2 type adjustment knobs, which adds to the weight, and confusion. Secondary sights shouldn't need to be adjusted once they are set at the yardage you feel is important. If an A2 drum is exposed to the daily, harsh treatment of the average soldier or law officer, it will not stay zeroed for long. I prefer a rear BUIS with an A1 type adjustment wheel to eliminate this problem.

Elevation Adjustments

You may also notice sights that give you elevation adjustments for extreme ranges; great idea if you plan to shoot extended ranges and not mount another optical sighting system on your AR. However if you are in the market for a back up sight, this may not be the answer. I have noticed that my elevation adjustable sights have been moved after allowing others to check out the type of sights on my rifle. This could have a catastrophic outcome. At the very least you should always mark your sights with a paint pen to insure you know where your sights should be set for the desired zero.

ALTERNATE BACK UP SIGHTS

You may not want BUIS (back up iron sights) only for when you have a broken scope or dead batteries. There may be a need to have a sighting system for close range that can be used in conjunction with your scope.

JPoint

The JPoint can be used as a short range sight when it is mounted on the top of a tube scope, or, with a special adapter, it can be easily mounted on the top of a Trijicon ACOG. See FIGURE 3.16 ACOG with JPoint. This setup will not allow you to establish a good stock to cheek weld when firing the weapon, but it is still handy for room distance engagements.

Iron Sights

Iron sights can also be mounted in such a way that they will be visible when using your powered scope. The Trijicon ACOG has just such a set of sights; they are not especially easy to use, but they may help for close range engagements. If you train, you will become proficient.

How often do scopes fail?

In today's modern age of sighting systems, the scope failure rate is minimal. I make this reference to optics that are not red dot sights. Red dot sights do fail, but if you have your BUIS readily available, you will be able to "Stay in the Fight." I have rarely seen any magnifying optic fail. One exception, of course, is if the sight gets hit with fragmentation or a bullet. See Figure 3.17. Modern scopes are made extremely well. Their rate of failure is so insignificant that I do not pay it any mind. I will caveat this statement by saying the scope must be one of the quality scopes I listed above.

Figure 3.17 ACOG that was shot

Figure 3.18 LaRue Tactical Mount

Mounting your scope

Mounting your scope can be a test of your patience, maybe not at the time of mounting; however, when you get to the range, carelessness will show with crocked or loose scopes. There are easy ways to avoid this headache. First, always use solid mounting systems, screw or throw lever systems. Do not use any accessory that is designed to be removed with your fingers. Other than the throw lever system, none have proven themselves worthy on a combat carbine. If you are planning to use any screws in your system, which is unavoidable, try to always at a minimum use blue Loc-Tite; red works too, but can cause some issues if you ever decide to remove the scope or rings. If you use Loc-tite, ensure the system is mounted where you want before it sets up. Once the thread locker sets, you won't be moving anything without starting all over.

If you are using a scope with separate rings, such as the LaRue tactical mount shown in Figure 3.18, you will need to also make absolutely sure the scope is mounted straight with the world (or at least straight with your upper receiver). To do this, you can use a small level on the receiver's flat surface, with your rifle secured loosely in a vise. When it is aligned, mount your scope and align the bubble on the scope as well. If you have a front sight block, do not use it as a reference, it may be slightly off. Always use the upper receiver as a guide. See Figure 3.19.

Figure 3.19 Level on receiver

This process is simple and normally works well. Another product that can help is the Reticle Leveler. This gadget is easy to use and very quick if you don't have access to a level and a vise. See figure 3.20 Reticle Leveler. It can be tricky but with a little practice you can easily figure out if you are aligned with your scope.

When setting up the eye relief on your scope (eye relief being how far away your scope is positioned from your eye to ensure a complete field of view without shadowing), always test eye relief in several positions. If it works well in a standing position, it may not work well in prone or kneeling.

Figure 3.20 Reticle Leveler

As I said earlier, check it before you head to the range. It will save you a lot of money on Tylenol.

Since this weapon is intended for tactical use, make absolutely sure you can still access all controls on the rifle. If the scope will not allow you to get to the charging handle, for example, you may need to modify the rifle or move the scope to allow access to these controls.

Tricks of the trade

When shooting your Red Dot for zero, turn the dot down to the lowest setting with which you are still able to see the dot on your target. This will help you to refine your hold on the target which will ultimately lead to a better zero on your rifle.

Choose the right sights for your application

When you are deciding which scope will best suit your needs, don't go to the latest issue of Gun Nut magazine and buy the cover scope. Ask around. See what some of the better tactical shooters are using. There are so many different ways to outfit your rifle with optics that you may end up with two options. One for certain applications and another on a LaRue quick detach mount for changing scenarios.

4. HEADING TO THE RANGE
Start With The Basics

It doesn't matter what your ultimate application for the AR system is, it is imperative that you familiarize yourself with the weapon before you use it in combat. Spending time on the range is an integral part of learning how to most effectively use the weapon system that you have chosen. When mastered, the AR can be a very effective weapon system.

Practice may not make perfect, but it will help to develop the reflexive weapon handling skills that will help you to shoot accurately in a hi-stress situation. Range and dry fire training are essential to developing a high level of confidence with your AR; and in a combat situation, confidence is key. An engagement is not the time to learn about the finer points of weapon manipulation. It is essential that you are familiar with your weapon before it is time to chamber a bullet and use it.

There are a variety of skills that you should practice before heading to the range. In later chapters I will discuss some skills that should be practiced and mastered during the course of several dry fire sessions. I will talk about loading your AR as well as positive check techniques to ensure the rifle is loaded. I will also describe speed and tactical reloads, as well as proper techniques for unloading or clearing the weapon.

While practicing these techniques, it is important to always attempt to keep your firing hand on the pistol grip. This will allow better control of the weapon and control of the weapon is vital if you plan to use it quickly. Positioning your hand on the pistol grip also allows the shooter to be aware of the muzzle of the gun during stressful conditions.

Loading /Reloading the AR

You are on the range, your heart races as you breath in. You squeeze the trigger, anticipating a bang and recoil of the weapon; you are instead surprised to hear a "click." How many times have you been on the range and heard you or your buddy fire one round and then encountered this embarrassing scenario?

This situation is the result of poor loading technique. In the following paragraphs I will discuss methods for properly loading the AR. If you follow these steps, you should not experience the one shot and click or the "one shot and dropped magazine" embarrassment.

First, place the weapon on safe and point it down range or in a safe direction. Lock the bolt to the rear by pulling the charging handle to the rear and pressing the non paddle portion of the bolt catch, the bottom portion. This action will lock the bolt to the rear. Next, push the charging handle back to its original position and be sure that it locks. See Figure 4.1.A-C.

Figure 4.1.A Locking Bolt to the rear

Figure 4.1.B, holding the charging handle back

Figure 4.1.C, pushing charging handle forward

Before placing the magazine in the weapon, be sure to check that it is fully loaded. This can be accomplished by pushing down on the top round in the magazine. See Figure 4.2A-C The first round should be on the top right side of the magazine. This will be true whether you are loading 28 or 30 rounds. Magazine specifics will be covered more thoroughly in Chapter 5.

Figure 4.2A Fully Loaded Magazine

Figure 4.2B Pushing down on top round of magazine 30 Rounds

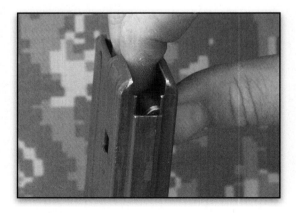

Figure 4.2C Pushing down on top round of magazine 28 Rounds

Place the magazine in the weapon and seat it until it clicks. Pull down to ensure the magazine is secured with the magazine catch. See Figure 4.3 Release the bolt by slapping the paddle portion of the bolt catch, letting it slam into battery. Watch the ejection port as you do this. See Figure 4.4A.

There are other techniques that can be used to accomplish the same result. My preferred technique is to let the charging handle go forward by hitting the bolt catch release with your thumb. See Figure 4.4B. Also, you can release the bolt by grabbing the charging handle and pulling to the rear until the bolt is released. See Figure 4.4C. You should see the round going into the chamber. Close the dust cover. Do NOT press check. A press check is when you slightly pull the bolt to the rear with the charging handle, just enough to visually inspect the chamber to see if there is a round loaded. See Figure 4.5. This technique is used extensively in the Military and Law Enforcement communities but just because it is used, it doesn't' mean it is right. A press check is the easiest way to induce stoppage before you even get started. If the weapon is dirty from the elements (i.e. sand from Helo operations) or has any carbon build up, you are substantially increasing your odds of a malfunction. The alternative techniques I will suggest will work with weapons that do not have a forward assist, thus developing muscle memory that can be used with several different firearms.

Figure 4.3 Pulling down on Magazine to ensure it is secured

Figure 4.4A Releasing the bolt catch by slapping the the paddle with your palm

Figure 4.4B Releasing the bolt catch with your thumb

Figure 4.4C Releasing the bolt catch by pulling the charging handle to the rear and releasing

Figure 4.5 Conducting a press check

To properly check for a loaded AR you must first push the magazine release and grab the magazine with your non firing hand. Place your index finger over the top of the magazine to feel if a round was stripped into the chamber. See Figure 4.6. If your magazine was full to begin with, the first round should now be on the left side. This technique will work day or night with flight gloves or mittens. Once you are sure a round has been loaded, place the magazine back in the weapon. Push it home until you feel the magazine catch and click. Now pull down again to ensure the magazine is locked securely in place. This last step is very significant. If you do not pull the magazine to ensure a secure lock, you will hear the disturbingly loud click of the hammer on an empty chamber as the magazine will immediately fall from the weapon upon firing if it is not seated properly.

Figure 4.6 Checking top round of magazine with index finger

Speed Reloads

When it is time to conduct a speed reload, you can use a modified version of your initial loading technique. For speed reloads, the bolt should be locked to the rear if your weapon is functioning properly. First, drop the empty magazine, See Figure 4.7. Let it hit the ground. If you successfully load, reengage the bad guy, and win, then there will be plenty of time to pick up the empty magazine. Seat a fresh magazine until it stops. Pull down on the magazine to ensure that the magazine is seated. Releasing the bolt can be accomplished by either of two methods. The technique I use is to press the bolt release with my thumb after seating the magazine; See Figure 4.8. Some use the palm of their hand. See Figure 4.9. Either of these techniques will work but require more training than the following, fail safe technique. The preferred technique for tactical shooters who will not be able to train extensively, is to simply grab the charging handle and pull it to the rear. See Figure 4.10. Once it has been pulled, release the charging handle allowing the bolt to slam in to battery, as it would during the normal firing sequence. If you do not need to fire immediately, once again repeat the check procedure outlined above.

Figure 4.7 Dropping magazine

Figure 4.8 Releasing bolt catch with thumb

Figure 4.9 Releasing bolt catch with palm

Figure 4.10 Releasing bolt by pulling the charging handle to the rear

Figure 4.11A Gripping the magazine as you would a pistol magazine, note magazine in carrier with bullets facing to the front

Figure 4.11B Pistol type magazine grip

Figure 4.11C Loading rifle with pistol technique

Figure 4.11D Pistol-type technique not conducive to power and does not allow the shooter to easily pull down on the magazine

Grasping the Magazine. When grasping the magazine, there are several ways to accomplish the task. But how can we build in a technique that will ensure success? For years I loaded my carbine magazines with the same technique that I use for a pistol. See Figure 4.11A-D This technique may work with 20 round magazines, but does not work well with 30 rounders. Rather than grab the magazine with this hold, you will have more power to seat the magazine if you hold as shown in Figure 4.12A-E. Not only does this allow you to seat the magazine with more power, but it also allows you to be in a better position to pull down on the magazine to ensure that the magazine is secured.

Figure 4.12A Correct grip on magazine, note magazine is placed in carrier with bullets to the rear

Figure 4.12B Correct magazine grip

Figure 4.12C Correct magazine grip, note pinky finger under bottom of the magazine

Figure 4.12D Much more power to seat the magazine and the hand is in place to pull down on the magazine to ensure it is seated

Figure 4.12E View from the opposite side of the weapon

Tactical Reloads

If you feel that you need to top off your weapon before moving to another position or after you have neutralized a situation, do so as follows: Find a secure location that provides cover as you reload. Drop the partial magazine into a pocket or magazine pouch that will not be used for hasty loads; you do not want to accidentally pull out a half loaded or empty magazine in a stressful situation. Locate a magazine that you know is full and to be positive, feel the first round to see which side of the magazine the round is positioned. The reason behind this check is to positively identify which side the first round is on. It is common that when a magazine is carried around in a pouch for an extended period of time, a round may be lost. Once the magazine has been placed in the weapon and you have pulled down to check for seating, you should be good to go. If you have any doubt as to the status of your weapon, pull the bolt to the rear and let it slam forward. This will give you confidence that the weapon wasn't already dry, or did not pick up a round from the previous magazine. This is simply to make sure that the weapon will be ready for action when it arises. Now pull the magazine from the weapon again to conduct your load check. If your particular situation allows, recover the round you ejected while loading.

Unloading

Once you are finished at the range or are in a secure location that may require a clear weapon, it is time to unload or clear your weapon. I have heard several stories of Accidental Discharges. They often occur as a result of shooters improperly unloading and clearing their weapon. To eliminate this issue follow these procedures.

Ensure your rifle is pointed in a safe direction and the safety is engaged. Remove the magazine with your non firing hand. Place the magazine in your drop pouch or pocket, the same location you would place the magazine during a tactical reload.

Once the magazine is stowed, pull the charging handle to the rear. As you pull the charging handle to the rear, watch the ejection port to insure you see the chambered round ejected. See Figure 4.13. Once you have pulled the bolt to the rear, visually inspect the chamber to ensure the weapon is in fact clear. See Figure 4.14. When you are on the range, and have visually inspected the weapon, point the weapon in a safe direction (down range at a target), place the weapon on fire, and drop the hammer. If you plan to store your weapon, close the dust cover and place it away. If you plan to continue carrying the weapon, pull the bolt to the rear and release. This will allow you to once again place the weapon on safe, now close the dust cover.

Figure 4.13 Round ejecting

Figure 4.14 Inspect Chamber

These procedures may seem mundane while you are practicing, or they may feel out of place if you normally conduct press checks. If you practice until they become reflexive, you will be able to confidently load, reload, and unload. Practice will also make you more comfortable when you have to complete any of these tasks at night. Build confidence in every aspect of your weapon handling; it will save your life.

5. MALFUNCTIONS
Magazine Problems

It doesn't matter where I go, every time that I conduct a carbine course it is inevitable that I will see magazines that look like they spent more time in Vietnam than Jane Fonda. I would recommend that you save up a few months to buy quality AR magazines before you spend a minimum of $650.00 on a quality, complete, AR variant.

Being a U.S. serviceman, I take weapon reliability very serious. I would never endanger one of my mates because I wasn't willing to spend a few extra dollars on a reliable magazine. When I head to the range, I always insure all magazines are marked with a number as well as an identifying mark such as my initials. See Figure 5.1 This procedure is not to keep some Range Thief from stealing my magazines. It is to allow me to keep record of which magazines cause problems.

Are all malfunctions caused by your magazines? No! Absolutely not! However, a large majority are. This procedure will help you quickly sort the wheat from the chaff.

I will describe several key indicators of poor or worn magazines. If you determine you have a bad magazine, you need to destroy the magazine. That's right, destroy it. I don't give it away or sell it at the Gun Show. I crush the magazine so no one can use it. I would not want to live with the burden of a friend or fellow serviceman losing his life due to one of my bad magazines. If you do decide to keep a few bad magazines around for malfunction drills, just insure they are marked. Paint them red, whatever it takes to clearly indicate their status. See Figure 5.3.

If you are loading your magazine and the rounds pop back out, this will indicate worn feed lips. Don't waste your time trying to reform these lips. Once the lips are bent, they will never be the same.

Figure 5.1 Magazine marking

Figure 5.2A Magazine with worn feed lips/compared to good magazine, Bad Magazine on the left

Figure 5.2B Note more exposed cartridge on the bad magazine on the left

Figure 5.3 Magazine marked for inducing malfunctions

If the magazine will not drop free when empty, I get rid of it. This problem will come with years of dropping the magazines on concrete or stepping on them. You may be experiencing spreading feed lips, if this is the case, there will be more problems than the magazine not dropping free. Bent feed lips will sometimes cause double feeds. If this happens, the easiest way that I have found to check the magazines will be to first visually inspect for bent or cracked feed lips. You may see a small crack on the rear of the magazine, where the magazine starts its bend. Also, you can download the suspected magazine to approximately 20 rounds. Next, tap the bottom of the magazine with the magazine held vertically. If the feed lips are bent enough to induce a stoppage, they will also allow a round or two to pop out of the magazine when this test is performed. See Figure 5.5A-B.

Figure 5.5A Conducting tap test on the bottom of magazine, checking for spread or bent feed lips

Figure 5.5B Magazine has been tapped, note round that has ejected from the magazine

If your 30 round magazines will not hold 30 rounds, they need to go in the throw away pile as well. I have heard a lot of shooters say "never load more than 28 rounds." Wrong! Quality 30 round magazines will work when completely full. If you do not feel comfortable loading with a 30 rounder due to difficulty seating the magazine, then go with 28. Just be careful to always check that they are loaded with 28 and not downloaded beyond 28 rounds. This check should be performed every time you initially load your weapon. You can easily check this by pushing down on the top, right round. See Figure 5.6. Push until you feel the rounds stop moving, make a mental note of this depth so you will be able to check night or day, without looking at the magazine. I personally always load 30 rounds in my magazines. I do not have difficulty making the initial load or any subsequent loads. Most soldiers I work with also follow this mindset.

Figure 5.6 Checking for fully loaded magazine

Another test I conduct with my magazines is to place them on the ground while shooting prone. I push forward with my shoulder to apply rearward pressure to the front of the magazine. I know, I know, this is another Urban Legend. Get over it and use the magazine for support whenever you need to. Using it will allow you to be a lot more stable and allow quicker follow up shots.

Magazine Choices

Which magazines should you buy? It's not too difficult of a choice, simply buy magazines that work. If you are cheap or, shall I say, thrifty, this may be difficult for you. If you are willing to suck it up and buy good magazines, it will save you money in the long run. There is no-one magazine manufacturer that I would recommend. When these magazines are made, you may get magazines from a good batch or you may get magazines from a bad batch. The only way to find out which batch the magazines came from is to load them up and test fire with these magazines. They may be fine initially, but later down the road you may start to have issues. When issues begin, it will indicate it is time to buy more magazines to replace the worn ones.

There are a couple of checks you can make to confirm you are purchasing quality magazines. First of all, is the magazine follower black or light blue/green? If it is black, avoid the magazine. If you decide to replace followers with after market followers, such as the Magpul followers, make sure the rest of the magazine components are quality as well. See Figure 5.8 Magpul follower.

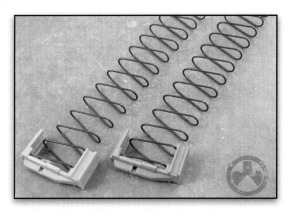

Figure 5.8 Magpul follower, courtesy Magpul

Insure that you have magazines that will drop free from your rifle. If they won't drop, get rid of them. If they have bent feed lips, get rid of them. Let's just say that it's best if you don't get too emotionally attached to your AR magazines.

Ammunition Choices

I cannot recommend one particular ammunition choice for the AR. There are so many different applications that the AR will fulfill, that it makes this choice almost impossible. I can, however, give you a well defined list of prerequisites that the ammunition should meet before you use it in your tactical AR.

First and foremost in my book of ammunition requirements is uncompromising reliability. I place this above my accuracy requirement. If your rifle will not function, who cares how accurate it is? When ammunition is built to Military Specifications it may not be cutting edge as far as accuracy is concerned; it will, however, be extremely reliable. I will caveat this with one statement: when I say Mil-Spec, I do mean for the US Military. There are several reasons Mil-Spec ammo performs when others cannot. The list of specifications that manufacturers must abide by when offering ammo to the military is much too long for this document. I will touch on the points that I believe contribute to the outstanding reliability of Mil-Spec ammo.

First I will address the brass used for Mil-Spec ammo. It must meet certain specs guaranteeing extremely hard brass that is not brittle. If brass is too soft, it cannot withstand the rigors of being jerked from the chamber in the manner that the AR requires. I have used brass that was not up to the task and this ammo resulted in stuck cases as well as cases with the rims torn completely off. Either of these cases could be catastrophic in a combat environment. The solidity of the brass not only contributes to ease of quick extraction, but it also helps to avoid stuck cartridges in a dirty chamber. Although these two benefits are enough to convince me to use the brass, I will continue to describe other benefits of Mil-Spec ammo. Not only is the brass hardened, but it is annealed too. Annealing makes the neck of the cartridge a different hardness, something that also contributes to reliability. The hardness of military brass changes from the front to the rear of the case.

> **ANNEALING:** Annealing is often thought of as heat treating for brass. Some people think that annealing is some sort of voodoo reserved for accuracy crazed hoodlums. Heat treating brass, or annealing, will cause different effects when applied to brass. Brass will not become harder when quenched in water after having heat applied. It will actually become softer. Brass cartridges that have been annealed in the neck area of the case will be a lot more consistent, thus giving better accuracy and better performance. To see if your brass has been annealed, look at the neck of the case. The neck of the case should have a rainbow discoloration in the neck area. This process is only applied to the neck. If the whole case was annealed, the case would not hold up throughout the firing sequence. Mil-spec cartridges are normally annealed; this only helps to add to their consistency from lot to lot.

Next are the specs required for the primer in this cartridge. These primers are crimped as well as glued into the primer pocket. This is not needed if you are planning to go hunt prairie dogs. If your life is on the line, it could keep you among the land of the living. When a weapon is fired on full auto, or as you fire semi auto, if the primers are not sufficiently secured in their pocket, they can be blown out or vibrate loose causing a malfunction with that round. The loose, floating primers can affect your trigger group as well as become lodged in the locking lugs of the rifle. If the primer ends up in the locking lugs, not only will the weapon not fire, but you may have a hard time getting this primer out of the way and getting back in the fight. Because of this spec, military brass is not the best for reloading. You must swage the primer pockets before the cartridges can be reloaded.

Not only are the primers held in the case with a little extra special something, the bullets are held in as well. When mil-spec ammo is loaded, the bullet used has a cannelure into which the mouth of the case is crimped. See Figure 5.9 Bullet with cannelure. This crimping also helps to degrade accuracy; however, if our primary concern is reliability, this is an acceptable sacrifice. In the future, special glues will be used to replace not only the cannelure, but the crimp as well.

Figure 5.9 Bullet with cannelure

You will have to determine which bullet weight and type of bullet construction meets your tactical needs. Once this is determined, find out which manufacturers offer acceptable loadings for tactical use.

Malfunction Clearance

Once you have determined that you have good magazines, a properly lubed action, and quality ammunition, it is time to practice malfunction drills with your rifle. First of all, if you are carrying a sidearm, you will normally just transition to that weapon if your primary weapon malfunctions. You should always practice malfunction drills with your rifle, just in case grabbing your pistol isn't the answer; for example, if your threat target is outside normal sidearm engagement ranges or if you have only a small portion of the target that is visible to engage. This shot may require the accuracy or reach that your rifle brings to the fight.

Normally shooters do not receive enough training on malfunctions. I continually drill students to ingrain a second nature reaction. Manipulation drills, such as these malfunction drills, help to build confidence in the shooters ability to handle their rifle safely and quickly in a stressful situation.

There are several different malfunctions that can occur with the AR. However, two basic techniques will get you through most malfunctions: Immediate Action and Remedial Action.

Immediate Action

In the military most instructors teach SPORTS (**S**lap, **P**ull, **O**bserve, **R**elease, **T**ap, **S**queeze). Slap the bottom of the magazine, Pull the charging handle to the rear, Observe the ejection port for obstructions, Release the bolt letting it slam forward, Tap the forward assist to insure the bolt is properly seated, and lastly Squeeze the trigger to hear it go bang.

I teach SPORTS first because it is truly the most effective SURE way to fix a simple failure to fire. The only difference between SPORTS and TAP, RACK, BANG (see next section) is the tapping of the forward assist and the "observing" the ejection of the case/round. I emphasize the "observe" portion, because as you increase your experience, you will recognize if there is a problem without consciously "observing." I call that the "jumpmaster theory" once I know what "right" looks and feels like, identifying wrong is simple. Also, if I am conducting immediate action, this means my rifle just didn't work when I needed it, so I want to take every possible action to ensure it will fire the next time I press the trigger. Eventually, it becomes Tap-Rack-Bang, but that comes with experience. I think if you start out with SPORTS, and then let it migrate to Tap-Rack-Bang, you are building a solid foundation.

This technique is very basic, however it does teach the soldier to analyze the malfunction. As the soldier becomes more experienced, the technique will evolve into Tap, Rack, Bang.

Tap, Rack, Bang

Immediate action should put you immediately back in the fight or allow you to find out if you have bigger problems than you were hoping for. I rely on Tap, Rack, Bang, to quickly get my rifle ready to fire. This is the same procedure used for pistol malfunction drills. Upon realizing you have had a malfunction, immediately tap the bottom of the magazine. Ensure that your trigger finger is outside of the trigger guard when you begin this process. This must be very assertive. A soft tap will not accomplish what is needed. You are tapping to either completely seat a magazine that was not properly seated or to loosen the stuck cartridges in the magazine, allowing them to smoothly feed to the top of the magazine. This tap should be applied with the left hand for right handed shooters, allowing the shooter to maintain a shooting grip on the pistol grip of the weapon. Left handed shooters should also use their left hand to tap the magazine. This somewhat awkward position is quicker for southpaws than using their non firing hand. This is due to the charging handle only having a release on the left hand side of the weapon. You are giving away your firing hand grip, but it is quicker, much quicker. See Figure 5.10A-5.10J.

Once you have tapped the bottom of the magazine, it may be necessary to tuck the butt stock of the weapon under your firing shoulder. See Figure 5.11. This technique will give you leverage to allow you to hold the weapon with your strong side hand only. When you conduct this sliding movement, do not raise your muzzle. Try to keep the muzzle relatively level, aiding in quickly re engaging the threat target once the malfunction is cleared.

This isn't especially necessary for left handed shooters since they will be leaving their forward hand in position. The center of balance should fall somewhere close to where their non firing hand is located.

Next, with your finger still off the trigger, pull the charging handle fully to the rear as you roll the weapon slightly to the right. This action will allow any loose cartridges or debris to fall from the ejection port. Release the charging handle and let it fly into battery. Do not ride, or ease, the charging handle forward. Easing the charging handle forward has high potential for not clearing the existing malfunction. You may be inducing another stoppage, which is the bolt not going completely into battery, or you could quite possibly be exacerbating the existing malfunction. If you are a right handed shooter you should be able to pull the charging handle to the rear with your left hand. This can be accomplished by simply grasping the charging handle release in the crook of your left index finger and pulling quickly to the rear. You do not need to grasp both sides of the charging handle to be successful. This technique is extremely quick.

Figure 5.10C RACK the charging handle to the rear

Figure 5.10A If weapon does not fire conduct Immediate Action.

Figure 5.10D

Figure 5.10B TAP the bottom of the magazine with as much power as you can

Figure 5.10E When releasing the charging handle, ensure that you keep your hand clear

Figure 5.10F BANG Reassess the threat, if it still exists, fire the weapon.

Figure 5.10I Left Handed. RACK the charging handle to the rearr

Figure 5.10G Left Handed. If the weapon does not fire, conduct Immediate Action.

Figure 5.10J BANG Left Handed. Reassess the threat, if it still exists, fire the weapon.

Figure 5.10H Left Handed. TAP the bottom of the magazine.

Figure 5.11 Tucking buttstock under arm to allow easier malfunction clearance

If you have a receiver plate mounted to your rifle for attaching a sling, ensure that there is enough room to accomplish this task. There are sling attachment points that impede the shooter from being able to quickly conduct immediate action. These types of devices should not be used. See Figure 5.12 Receiver Sling Attachment.

Figure 5.12 Receiver Sling Attachment

To eliminate the possibility of riding the bolt forward, I pull the charging handle to the rear until my left finger slips off the side of the charging handle. Once the bolt has been released forward, re assess the target. Ensure that you still have a threat to engage. Re grasp your weapon and attempt to fire. If the weapon will not fire you will now move on to Remedial Action.

While clearing malfunctions, you should always attempt to do all the work with your non firing side. This is true for right handed shooters. Left handed shooters need to see what your weapon will allow you to get away with. Some shooters have added ambidextrous controls to their weapons. These controls make manipulation much easier for left handed shooters; however, left handed shooters should also practice alternate means of accomplishing these tasks. What will you do if you end up with a rifle that doesn't have all the bells and whistles?

Remember, as you are tapping the magazine you should let your eyes quickly assess the ejection port. This will help you to quickly analyze the malfunction.

Remedial Action

If Tap, Rack, Bang, does not work it is time to move on to Remedial Action. During Immediate Action your goal was to quickly eliminate the stoppage because you have an imminent threat. Nothing has changed. You will still have to decide if the threat is imminent and figure out how you can you deny him the chance to take your life. Are you going to run over him, kick him, or use the old standby, scratch his eyes out?

If the threat is not in close proximity and not delivering effective fire, you will have more time to get the weapon system working again. You should still attempt to seek cover; since, at this time, you do not know the severity of your malfunction. In most cases Remedial Action will require removing the magazine. See Figure 5.13. As we discussed earlier, magazines are the cause of a large percentage of malfunctions; therefore, do not reload the weapon with the same magazine. After the magazine is removed from the weapon, attempt to lock the bolt to the rear. See Figure 5.14. If the bolt will not lock to the rear you may have to cradle the weapon in the crook of your left arm, using your thumb on the left hand to hold the bolt as far to the rear as possible. See Figure 5.15. Or hook the bolt with your firing index finger for right handed shooters. See Figure 5.16. As a last resort use the sharp corner of your magazine to pull the bolt to the rear if you don't have the big, rubber, loops on your magazines. See Figure 5.17. You shouldn't need to use a knife or Leatherman tool to clear most malfunctions.

Figure 5.15 Cradling weapon while pushing bolt to the rear with thumb of support hand

Figure 5.16 Pulling bolt to the rear with index finger

Figure 5.13 Removing Magazine

Figure 5.17 Pulling bolt to the rear with magazine

Figure 5.14 Locking Bolt to the rear

Once the bolt is pulled to the rear, use your fingers to pull the jammed rounds from the feed ramp. See Figure 5.18A-5.18C You should be able to get a minimum of one of the rounds out of the bolts way to allow forward movement and extraction/ejection of the stuck cartridge. When using my fingers, I normally go from the bottom of the weapon through the magazine well. If your bolt is stuck forward on a spent case, please refer to the Stuck Case section of this chapter. It will talk in depth about how to rectify the problem.

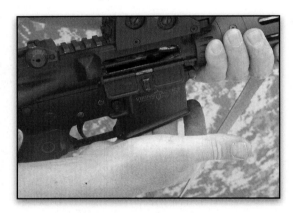

Figure 5.18A Using fingers to reach from bottom of magazine well and push rounds loose.

Figure 5.18B Continuing to push rounds to get double feed to come loose.

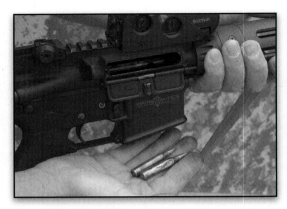

Figure 5.18C Double feed rounds should fall out into hand.

Analyzing Malfunctions

During your range training sessions you do not always have to clear your malfunctions in a speedy fashion. You may want to take a minute to analyze this situation. This will save you time in the long run. You should become familiar with what these malfunctions are going to look and feel like. It is also important that you figure out what the source of the problem may be. Get used to what it feels like to shoot your weapon dry. If you recognize this feeling, then you can immediately reload or transition to your sidearm. This will definitely save time.

> *Shooting your weapon until it is out of ammo or "dry," as we call it, is important. It is important to learn what the rifle feels like when it is fired untill it is empty. If the bolt locks to the rear during the firing sequence, you should immediately know that you have run out of ammo. It has a certain feel that is only learned by spending.*

time on the range. This will make your magazine changes faster or your transition to your sidearm appear as though you were cheating.

Double Feed

The Double Feed is two live rounds in the feedway of your weapon system. This malfunction is often confused or misnamed with a failure to eject. This may lead to misdiagnosis of the problem. The Double Feed is caused by a poor magazine. Double feeds are normally not very difficult to clear unless you have aggravated the situation by slamming the bolt forward. Often, undue beating and banging of the weapon system will only serve to make the situation worse. If you wonder what it will look like when you have this malfunction, you will have two live rounds attempting to insert into the chamber at the same time.

This stoppage can normally be quickly cleared if you have a bit of patience. We covered the double feed earlier in this chapter but will revisit the topic now. Double feeds will require Remedial Action. Lock the bolt to the rear, remove the magazine from the weapon, and hold the bolt catch so the bolt doesn't bite your fingertips. See Figure 5.29 Reach from the bottom of the magazine well to knock the cartridges loose. See Figure 5.30 You may need to simply tap the butt stock to get the rounds moving freely. See Figure 5.31 Other aggravated Double Feeds may require the use of a pocket tool to remove the stuck double feed. See Figure 5.32 Once you remove these cartridges, check them for damage. If there are any dents, severe scratches, or the bullet is bent in the case, get rid of it. If not, proceed to the instructions for a stuck case.

Figure 5.29 Holding Bolt catch

Figure 5.28 Double Feed

Figure 5.30 Reaching from bottom of receiver to release stuck rounds

Figure 5.31 Tapping buttstock to remove double feed

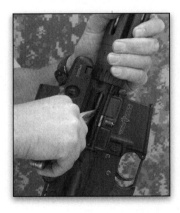

Figure 5.32 Using Leatherman to remove double feed

Failure to Extract/Stuck Case

A Failure to Extract is fairly simple to detect. If you pull the bolt to the rear and the fired brass comes out of the chamber, you had a failure to extract that is now fixed. This may have been caused by a poor extractor or a poor ejector spring.

Normally, if you have this type of failure it will not be cleared as easily with a Tap Rack Bang. It will normally be a more involved process.

This type of failure can be caused by several deficiencies. You may have a dirty or rough chamber that has stuck to the fired casing. If this is the case, clean and lube the weapon. If it persists, have a gunsmith polish your chamber, if your rifle has a chrome lined barrel this usually isn't the problem.

If it is not a dirty chamber, then it could be caused by a bent or deformed piece of brass. The cartridge could be so severely dented that it never made it all the way into the chamber. This can be caused at the factory or by poor care of ammunition. It can also happen when the bolt is allowed to slam into the side of a cartridge; such as during a bolt over base. See page 71. If you notice bent casings, dispose of them properly. Don't put these rounds in your magazine. You may also cause this malfunction by lubing your weapon too heavily. Excess lube can cause brass to get small indentations that will not allow the case to be pulled freely from the chamber.

Next, if the round is fired but does not extract, it may be a weak extractor spring or broken extractor. If you are able to pull the bolt to the rear without the brass coming free, pull the bolt to the rear, point the rifle in a safe direction, and gently tap the butt-stock to see if this round will fall free. If it does fall free, check your bolt and extractor for damage. If it doesn't fall free, you may need to get a cleaning rod to clear this malfunction. See Cleaning Rod Clearance Below.

If you have tried to use Immediate Action with no success, start the Remedial Action process. Drop the magazine, or pull it from the weapon. Failures to extract come in several types. The first being you have a sticky chamber. If you have a sticky chamber, allowing the bolt to slam forward onto the case and then grasping the charging handle and quickly jerking it to the rear may clear the case. You may need to un-sling the weapon and transfer your firing hand to the charging handle to attain a firm grasp. Collapse the butt stock and tap the butt-stock on your knee or the ground. See Figure 5.19A. If you attempt this drill and the charging handle will not move to the rear, you have ruled out a bad extractor. Try to tap slightly harder until the inertia aids in moving the bolt to the rear. If you grab the charging handle in your right hand, aggressively tap the butt stock against a wall or the ground to help dislodge the case. See Figure 5.19A-5.19D. Make sure to keep the muzzle pointed in a safe direction. If you are using a collapsible stock then it should be collapsed completely before attempting this clearance technique. If it does not extract after two or three tries, you may need to get out a cleaning rod to rectify this situation.

Figure 5.19A Collapse buttstock before attempting to dislodge stuck case.

Figure 5.19B Grab charging handle release and prepare to tap buttstock.

Figure 5.19C Tap buttstock on a solid object while pulling charging handle to the rear.

Figure 5.19D Clearing stuck case.

Make a note that there may be one of two reasons for this particular malfunction. If it was a stuck case and your extractor is not slipping off the case rim, you might be able to try one other technique. I use a Leatherman Wave for this fix. The reason I use the Wave is not only because it is a great tool, but because it has a slender nose on the pliers. This will allow me to reach from the bottom of the magazine well, place the nose of the pliers in the front of the bolt carrier, and then pry to the rear. See Figure 5.21A-5.21D. If this does not work, immediately go to your cleaning rod technique.

Figure 5.21A Leatherman Wave can be used to remove stuck case. But only if the bolt carrier can be moved slightly to the rear.

Figure 5.21B Slide nose of Leatherman Wave in front of bolt carrier.

Figure 5.21C As you pry the bolt carrier to the rear, the case will pull from the chamber.

Figure 5.21D If this technique does not work you will need to go to the cleaning rod technique.

Figure 5.22 Military Cleaning Rod sections

A basic military cleaning rod will work well to remove this stuck case. Slide the assembled cleaning rod down the barrel while pushing the rod to the rear and pull the charging handle to the rear. You may have to bounce the butt of the rifle against the ground a couple times to get the stuck case to break loose. Once the bolt starts to move to the rear, simply pull the cleaning rod out of the spent cartridge and the brass will eject from the weapon. See Figure 5.20A-F.

Figure 5.20A Cleaning rods carried already assembled and rubber banded to side of weapon.

Cleaning rods. Most tactical shooters do not carry cleaning rods on their rifle or in their kit. This can be a show stopper if you have a severally stuck case. The compact, cable, cleaning devices won't work for this emergency. What you will need is a good old fashioned, military issue, cleaning rod. I normally carry 4 sections of cleaning rod. This will clear not only my rifle, but someone else's who might be carrying a slightly longer, barreled rifle. Make sure that you remove the eyelet tip before attempting to use the rod for this duty. If you don't remove the eyelet tip, then it can bend and the rod will be lodged in the barrel. Let me just say, I learned this little tip from experience. See Figure 5.22

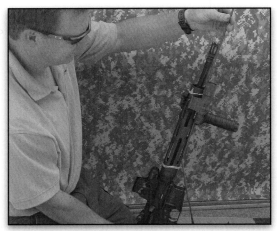
Figure 5.20B Removing cleaning rods.

Figure 5.20C Sliding cleaning rods down barrel.

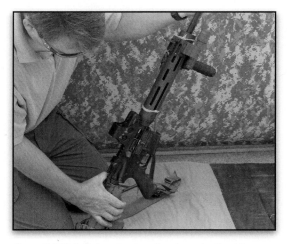
Figure 5.20D Pull charging handle release latch with thumb as you tap buttstock on the ground. You must also being putting pressure on the cleaning rod all the while.

Figure 5.20E Continue to push rod and casing into the upper receiver.

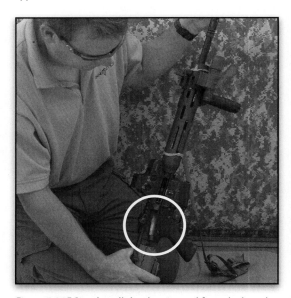
Figure 5.20F Simply pull the cleaning rod from the barrel and the case will pop out. Note brass ejecting.

If you are pulling the bolt to the rear with little to no resistance and you are sure your extractor is in good condition then look at the rear of the brass. You may have soft brass that has let the rim of the cartridge pull free. This can happen at any time, although it is very rare with Mil-Spec ammo. If you are unable to remove the stuck case, once again, get the cleaning rod. If the case has cracked in half and you are not able to see the front half of the casing, you may induce an even more severe stoppage when you attempt to let another round slam into the chamber. By the time you figure it out, you have more than likely attempted loading several times.

To remove the stuck, cracked, half piece of brass, you may need to take a chamber brush. See Figure 5.35A-C. Push the brush into the chamber as far as possible and attempt to pull the case out with the brush. If this does not work, take the rifle to a competent gunsmith.

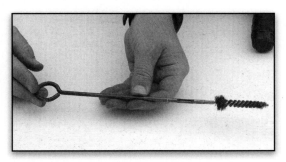

Figure 5.35A Cleaning rod with chamber brush.

Figure 5.35B Slide rod with brush from the rear of the weapon into the chamber.

Figure 5.35C Push the brush into the broken case and pull quickly to the rear to attempt to dislodge the split casing.

FAILURES TO FEED

Bolt Over Base

Bolt over base is not a class of malfunction, it is a description of a failure to feed, caused when the rifle bolt moves forward faster than the next round can feed from the magazine. If your weapon is only malfunctioning with one specific magazine and they are correctly numbered then you should make quick work with finding and fixing the malfunction issue. If your rounds are not being completely fed into the chamber, then it may be a magazine issue. This may be especially true if you have a bolt over base. Bolt over base is a condition that exists either due to a poor magazine spring, bent magazine, or your rifle is not getting enough gas to the system.

There are also cases where the bolt speed is too fast. This happens regularly when you use a sound suppressor on your weapon. Silencers create more back pressure. This back pressure increases the bolt velocity just enough to outrun the magazine. You may also have an issue with the bolt not locking to the rear on the last round fired. See Figure 5.23.

Figure 5.23 Bolt over base

Very rarely is ammo the issue if you are using name-brand, quality ammo. A bolt over base will have the spent cartridge successfully extracted and ejected; however, the next unfired cartridge will either try to be fed with the bolt over the lower half of the round or the round may not be fed at all, depending on the spring weakness.

To remedy this situation, conduct a Tap Rack Bang. If this does not work, move on to remedial action. Make sure you slightly roll the weapon toward the ejection port side to allow any debris to fall out of the port.

Replace the magazine. If this isn't the issue then you will need to try other ammunition or have a gunsmith check out your weapon. You should also check to makes sure that you have enough lubrication on the weapon system.

You will also notice that some of the brass from a bolt over base malfunction will have what looks like teeth marks in the side of the brass. See Figure 5.24.

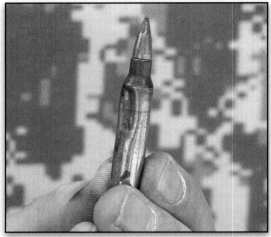

Figure 5.24 Dented Cartridge

There is always the possibility that the whole cartridge is slightly bent. If these cases are reloaded in order to try and shoot them again, you may be causing another, more severe, stoppage. If the bent case, now not a perfect cylinder, feeds into the chamber, then it will become stuck. If you try to force the case into the chamber, it will only make the situation worse.

Stove Pipe

A Stove Pipe, or failure to eject, does not necessarily look like a stovepipe in a pistol. Normally a stove pipe implies that you have a round sticking out like a stove pipe. See Figure 5.25A. As I said earlier, you will occasionally see this when using a pistol. With a rifle you may not have very much brass exposed from the rifle. If you have a large enough protrusion of brass, try to sweep it away, allowing the bolt to go forward into battery. See Figure 5.25B. If the bolt does not go far enough forward, then you may need to conduct a Tap Rack Bang.

Figure 5.25A Stove pipe

Figure 5.25B Clearing Stovepipe

Stove Pipes are usually caused by a weak extractor spring or a weak ejector spring. These parts are inexpensive and should be replaced periodically. If this situation persists, then it could be time to spend $2 dollars on these replacement springs to get the rifle working properly.

Figure 5.26 Extractor spring

On multiple occasions, I have seen a shooter take a position too close to cover, causing a malfunction. To avoid this situation, if you go prone to shoot under a vehicle, do not get too close to the tires. Getting close is great, unless it induces a weapon stoppage. I have seen ejected brass find its way back into the ejection port several times. Ole' Murphy is always there. You may also experience this problem when you try some of the positions in Chapter 10. Modify the positions slightly to allow enough clearance for the ejection port, permitting the weapon to function correctly.

If you have a weak ejector spring you may also notice small brass marks at the front of the ejection port or on the edge of your free float tube/delta ring. This is evidence that the rounds are barely clearing the port before the bolt starts its forward movement.

Additional Malfunctions

There are several other malfunctions that can occur. These malfunctions are more rare than those that I have already covered. When you have a malfunction, it is important to evaluate the problem and try to determine what the cause of the problem may be. If it is equipment related, obviously you should get it fixed. If it is ammo related, change the ammunition. There is a reason cheap ammo is cheap. If you don't understand why, you will quickly figure it out at the range.

Charging handle impingement malfunction

This is the most serious type of malfunction. It begins as a failure to properly extract, where the extractor loses control of the case at some point in the recoil cycle (as the bolt carrier moves to the rear). This causes the case to float away from the ejector, stopping it from expelling the empty case. The bolt carrier comes forward and feeds a fresh round from the magazine, which, while riding up the feed ramp, knocks the empty brass that did not eject up into the carrier key channel in the upper receiver. As the bolt continues forward, it partially feeds the live round into the chamber, while pinning the expended brass between the face of the bolt and the back of the charging handle claw (where the gas tube passes through).

This malfunction is easily cleared by pulling the bolt slightly to the rear with your right index finger. This will allow the spent brass to drop free from the lodged position. You will also need to slightly cant the rifle to the right to allow this empty case to fall free.

Bolt override malfunction (failure to eject)

This malfunction is much like the charging handle impingement malfunction. This happens for the same reason as the Charging Handle Impingement malfunction except the empty case is caught in the gap between the gas key, and the bolt and carrier. It is called a "bolt override" because the bolt has "overridden" a portion of the expended cartridge case. See Figure 5.36 Bolt Override Malfunction.

Bolt override malfunction (double-feed)

This malfunction is much like the failure to eject bolt override, but it involves two live rounds. This is most common when a soldier is firing on his side, as in under a vehicle. When the weapon discharges, the shock of the bolt carrier moving to the rear can dislodge a round from the feed lips of a weak or unserviceable magazine before the bolt comes forward. If the shooter is at an extreme angle, the live round can fall into the carrier key channel in the upper receiver. This can create a bolt override malfunction where the projectile portion of the live round is jammed in the gas tube.

Although these malfunctions are rare, you must know how to fix the problem. If this happens, it can be devastating if you don't know what to do to remedy the situation. You may also cause damage to the portion of your gas tube that extends into the upper receiver. If you find yourself in this predicament, attempt the following steps. See Figures 5.36-5.40.

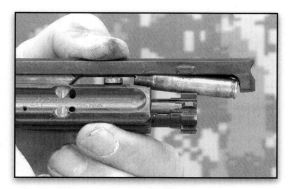

Figure 5.36 Case stuck between bolt carrier, bolt, and charging handle

Figure 5.37 Pulling bolt to the rear with finger

Figure 5.38 Pulling bolt to the rear with thumb

Figure 5.39 Pulling bolt to the rear with corner of magazine

Attempt to pull the bolt to the rear. If it will not move far enough to the rear to lock then you may need to pull the bolt with your finger, thumb, or the corner of your magazine. See Figures 5.37-5.39. As you pull the bolt carrier to the rear, ensure the charging handle is forward. You may not be able to do this until the bolt is locked all the way to the rear. Once it is locked, you will be able to give the charging handle a quick tap to put it back into position, hopefully freeing the brass. See Figure 5.40 If the brass is not freed, while holding the bolt catch in the locked position to ensure you do not crush your fingers if the bolt is accidentally released, reach in from the magazine well and attempt to free the stuck case. Normally, you never make it to this point because by the time you do this, the round will have already dislodged itself. It is also important that the muzzle of the weapon be pointed towards the ground to allow gravity to help clear the malfunction.

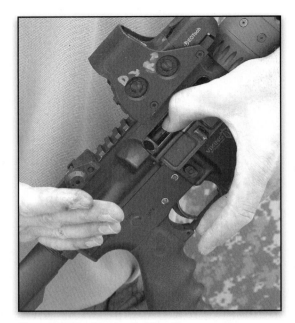

Figure 5.40 Tapping charging handle forward with bolt to the rear

Discussing all these malfunctions may lead you to believe that the AR is not a trustworthy system. Actually the truth is quite the opposite. This system is extremely reliable when proper ammunition, magazines and maintenance are provided. If you have a malfunction with the AR, do not accept it. Find the problem, fix it, and your malfunctions should be very few and very far between.

Dummy Rounds

If you have a rifle that has severe malfunction problems, get it fixed and never accept it. To train yourself to deal with unexpected malfunctions, use dummy rounds or training rounds. These rounds are marked in such a way as to allow the shooter to quickly realize which rounds are dummies. I prefer to use inexpensive, injection molded dummy rounds. These dummy rounds are fluorescent orange. See Figure 5.41 There are also rounds available that are nothing more than a real round of ammunition that has had a dummy primer and no powder added. I prefer rounds that stand out so that I can quickly find them on the ground when I am finished with the training. It is also useful in order to quickly identify when loading my magazines for real world use.

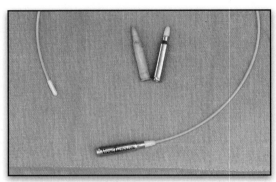

Figure 5.41 Orange Dummy rounds

If you go to your local gun shop, you should be able to find several types of dummy rounds for range usage.

When you head to the range, have your shooting partner load your magazines for you so you are somewhat surprised when the dummy rounds cause a malfunction. It is a little difficult to induce a double feed or a bolt over base, so have your partner call out malfunction commands as you go through your normal shooting drills. This will allow you to be evaluated by your buddy, as well as get familiar with quickly reacting without pausing to think about the correct procedure for reducing this stoppage.

Transitioning to Sidearm

If you have decided to carry a sidearm as a back up to your primary weapon, then this will allow you to immediately transition from a malfunctioning primary to a back up weapon system.

If you are carrying a sidearm for this purpose, then you need to practice these transitions. Practice them with a shooting partner so that you do not know when the weapon is going to fail or run out of ammunition.

If you are working by yourself, it is easy to load an odd number of rounds into each rifle magazine, mix them up, and fix the problem when it occurs. Conduct drills that require two rounds. This will require you to transition to successfully complete the drill. You can also buy dummy rounds for your rifle and mix them in with your practice loads. Just insure you check all magazines before leaving the range. You don't want to end up with dummy rounds when you need live ammunition. Once again, I recommend the inexpensive orange dummy rounds.

The reason I practice with downloaded magazines or dummy rounds is as follows: It is realistic. Some instructors require the students to place their weapons on safe while transitioning. This action will teach students the bad habits that are reminiscent of Police officers dumping spent revolver brass in their pocket. Either way you are going to end up taking a "dirt nap." If your weapon has malfunctioned and you squeeze the trigger and hear or feel a click, you will not be able to place the AR on safe. If you can, then there is something wrong with your trigger system. If the weapon has run out of ammunition, then you will feel the bolt lock to the rear (especially, if you train with the weapon frequently). If the weapon is dry, then you will not need to engage the safety either. Just make sure you are only conducting transitions when you go dry or have a malfunction.

I also notice students looking at their rifle before they transition. Don't look. If it doesn't work, then it is time to break Kydex or Leather or whatever your choice holster material might be. Get the pistol out as quickly as possible.

Once the engagement is complete, immediately work to rectify the issue with the primary weapon. You never know when you will need it again. It could be sooner than you think.

In a stressful environment (by this I mean gunfight, not coming home late from fishing), you will need to have these transition, loading and malfunction drills ingrained in you as second nature. You should also practice these drills with your support side, with one hand, with one hand on the support side, etc. You get the drift!! Be ready. "Murphy" is always there; we just have to outsmart him.

See Chapter 15B for step-by-step details on transitioning to your sidearm.

5B. TACTICAL SLINGS
Using the Tactical Sling

The sling is a tool that the Tactical Shooter will often overlook. With the exception of the Tactical Sniper, who must use a sling to be successful, most tactical shooters do not understand the true benefits of a good sling. Most shooters do not look at the sling as a multi-tool. When talking tactical, we cannot carry a tool for every job. We must try to find tools that cross over and provide multiple functions. In the past there were slings that were carry straps (AR sling), See Figure 5B.1 or shooting slings, but not both. The leather GI sling is a great sling. It can be used as a shooting sling or a carry strap; however, the transition time to convert the sling from one setup to another is not practical for the tactical carbine. See Figure 5B.3 Today, there are more options when looking for a working sling.

Figure 5B.1 AR Sling

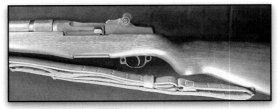

Figure 5B.3 Leather GI Sling

Slings on tactical rifles are normally employed as basic carrying straps. The military sling is designed for muzzle up, shoulder carry of the AR system. The sling attachment points on the AR do not lend well to the tactical shooter; however, with the evolution from the M16 to the M4, we have seen side sling swivels as well as collapsible stocks added, allowing proper sling placement for the tactical shooter. This not only allows the shooter to carry the weapon in a more readily accessible manner, but also allows for the use of the sling as a shooting aid.

I will cover several different tactical slings. I suggest you experiment with several sling types in order to allow you to choose the sling that best fits your application. I have had experience with the majority of the slings that I will discuss, and, in some cases, these are slings that I, along with my Army comrades, designed.

Three Point Slings

Three point slings or HK style slings are one of the best kept secrets in the shooting community. See Figure 5B.4. The only problem is that the secret happens to be that three point slings don't work well for tactical shooting. First of all, these slings cover the controls on one side of the rifle, and if you are left handed, they actually cover the ejection port. More is not always better. I have used several three point slings and still do not see the advantage of the extra material. Some companies have done a great job marketing their three point slings, but these slings do not aid the shooter in all situations across the tactical arena. Some shooters prefer three point slings for transitional purposes. Normally, they are only thinking of the transition to their sidearm. These slings work well for this task; however, they normally allow the weapon to sag too far from the body or they hold the weapon in a port arms position. If the weapon sags it will not be conductive to running or protecting the shooter from an extremely warm barrel. If the weapon remains in a port arms or muzzle up configuration, this is also unsafe due to the close proximity of other shooters in a tactical environment. It is unacceptable to cover a fellow shooter with your muzzle.

Figure 5B.4 3-Point Sling

Truly being able to transition a weapon during tactical scenarios is more than just dropping the weapon to grab your pistol if the primary weapon malfunctions. Your tactical transition may be the act of transitioning the weapon to your back for building climbing or apprehending a suspect. If this is the case, the three point sling does not work well to move the weapon to your back unless you have decided to sling the weapon muzzle up. Muzzle up carry of a tactical carbine should not be a technique used by experienced and well trained tactical shooters. If a medic were to sling the weapon muzzle up to treat a patient, he would be covering others with his muzzle every time he bent over. Clearly, this would be a major no go. Another transition you may encounter is removing a carbine from an injured mate. A three point sling makes this task somewhat more difficult when the extra strap of the three point sling cinches itself around the shooter. This can also affect a shooter who is not injured, but needs to remove his weapon quickly to conduct a car search or an attic search. The last transition of concern is the transition from strong side shoulder to support side shoulder. Most three point slings make this move difficult due to the cinching of the sling's extra strap around the shooters body.

The bottom line is that the evolution of tactical slings is taking us in a different direction these days. I think you will agree better solutions can be found.

Single Point Slings

Single point slings are creating quite a stir in the tactical shooting community. They are relatively simple and help the tactical shooter to have less interference with their weapon system than the three point sling. I would place the single point high on the list for a good carry strap. These slings also work relatively well during transitions. See Figure 5B.5A. Several single points also come equipped with bungee cord to allow muzzle strikes when using the rifle as a hand to hand weapon. I feel that I am able to strike just as well without the bungee, but different strokes for different folks. If you do decide to use a section of bungee in your sling, be careful when jumping over fences and remember that the bungee works both ways. I have seen several bloody lips from bungee slings.

The single point sling will aid your transitions to your sidearm as well as make it simple to sling the weapon to your side or back. You must be careful during transitions to the sidearm. The muzzle tends to drop straight down, rather than to the off side. See Figure 5B.5B. You do not have the ability to cinch the weapon tight across your back with most single point slings. You do have the ability to get it out of your way though. If you are planning to use a single point, I would also come up with a weapons retention device to hold the weapon in place for any type of building climbing or suspect apprehension. If you decide to attach the single point to a receiver plate on your rifle, make sure that this attachment does not interfere with proper function of the rifle, get in the way of your firing hand, or impede movements needed to clear malfunctions. See Figure 5B.6 Receiver plate. When clearing a malfunction with the non firing hand, does the attachment get in the way as you pull the charging handle to the rear? I will cover more specifics with regard to transitions from shoulder to shoulder with the single point sling in Chapter 12.

Figure 5B.5A Single Point Sling

Figure 5B.5B Transitioning to Sidearm with Single Point

Figure 5B.6 Receiver Plate

Two Point Slings

Near and dear to my heart is the two point sling. Once I realized a three point would not work well, I started to carry a simple two point sling. I even went so far as to just use a two quart canteen strap. Now that is the ultimate in practical! One of my buddies still uses a two quart strap. He is completely proficient and it works for him; but, like I said earlier, different strokes for different folks.

One of the down-falls of the two point sling is that you can not quickly adjust the sling. If the sling is being used as a carry strap and you decide to transition the weapon to your back, you are not able to adjust the sling to keep the weapon from banging and catching on other gear as you move. We remedied this predicament by making a quick adjust, two point sling. See Figure 5B.7 Viking Tactics Sling. It is extremely simple yet allows you to use the sling for every aspect of tactical shooting. As a carry strap it transports the weapon muzzle down, and can be adjusted quickly to longer or shorter adjustments depending on how you are wearing the sling.

Figure 5B.7 Viking Tactics Sling

Tightening the VTAC sling

Figure 5B.8A Normal carry position with VTAC sling.

Figure 5B.8B Grab free running end and start to pull.

Figure 5B.8D Pull until the sling is sufficiently tight.

Figure 5B.8C As you start to pull the weapon will be cinched against your body.

Figure 5B.8E With sling cinched tightly you are capable of climbing or laying hands on someone if necessary.

Releasing the VTAC sling

Figure 5B.8F Releasing the sling. Secure the pull lanyard with your fingers.

Figure 5B.8G As you pull the release lanyard the weight of the weapon will extend the sling.

Figure 5B.8H If desired you can let the sling all the way out to full extension.

Figure 5B.8I VTAC sling fully extended.

The quick adjust capability also allows you to lengthen the sling when transitioning the rifle to the support side shoulder for corner clears. The single point is slightly quicker for this task but cannot perform many of the functions of the VTAC two point sling. Another important reason you will need to adjust your sling rapidly is for target engagements. If you have a chance to get in the prone position to take a shot, then you should use your sling. With our quick adjust sling you can adjust to tighten your shooting position whatever it might be. See Figure 5B.9A-5B.9D. The two point also needs to be attached to the rear and top of the butt-stock, See Figure 5B.10, and near the front of the hand guard or near the front sight. See Figure 5B.11. This positioning will help to hold the rifle closer to your body during the shooting sequence. Earlier in Chapter 2 we talked about free float hand guard systems. If you plan to really crank on your sling as I do, then I would recommend a free float tube to eliminate the possibility of pulling rounds off of their intended target with too much sling tension. I have been able to move a group as much as four inches at 100 yards with extra sling tension.

Figure 5B.9B Grab free running end.

Figure 5B.9C Pull until the sling is tight.

Figure 5B.9D Engage the threat accurately with a tight sling, which will enhance your shooting position.

Figure 5B.9A Adjusting the VTAC sling while in the prone.

Figure 5B.10 Sling attached to top of buttstock

Figure 5B.11 Sling attached to side sling swivel

The complete adjustability of this sling also aids those of you who might have to fast rope or rappel. It will allow you to quickly cinch the weapon down during the act of roping, and then quickly release the sling to the proper length as soon as you hit the ground. This will also help as you move into a situation such as handcuffing or searching a suspect. See Figure 5B.12A-5B.12E.

Figure 5B.12B Reach up and grab end of VTAC sling.

Figure 5B.12C Start pulling to tighten.

Figure 5B.12A Weapon slung across back.

Figure 5B.12D Continue pulling until the sling is tight.

Figure 5B.12E With the sling tight across your back you are ready to fast rope or place handcuffs on one of your customers.

There are other two point slings on the market. One, in particular, is a copy of our sling with a slightly different adjustment buckle. This sling does not have a positive adjustment that locks and it does not allow you to continue to cinch the sling tighter and tighter. Once our sling is cinched tight, there will be a slight tail. I have never had any issues with this tail becoming caught or getting in my way. Give it a try and see what you think. I believe you will quickly become a fan of the VTAC sling. It should provide you the flexibility to perform any task across the tactical spectrum.

6. FUNDAMENTALS OF MARKSMANSHIP

The basic fundamentals of marksmanship apply no matter your shooting sport or tactical application. During combat shooting, we add one more element which may make the wheels fall of; that element is SPEED. If we are to be successful in a confrontation, then we must be quicker than our adversary. Bad guys rely on "Spray and Pray." We must rely on the basic fundamentals at a swift pace. Fundamentals are guidelines to be followed to cause the perfect end to the firing sequence. They are adjusted according to the circumstances of the target engagement. How close? Is the target moving? Is there a hostage obscuring part of the target? In combat shooting these fundamentals are not black and white; they are grey or, as some folks say, "they are as solid as JELL-O®."

Natural Point of Aim

Before we start with the Fundamentals of Marksmanship, let's discuss Natural Point of Aim. Natural Point of Aim is a term used to describe the ideal position your body should be in so as to maximize the position as well as to aid in the basic fundamentals of marksmanship. In layman's terms, it is a natural position that allows your body to relax, allows you to expend less energy keeping your sights on target, and aids in a quick, natural, return to target once the firing sequence is complete. The closer your position is to your Natural Point of Aim, the more durable your position will be. This will allow you to stay in the position comfortably while awaiting the opportunity to engage a target.

To find your natural point of aim in a prone position: Assume a comfortable, stable, yet durable prone position. Point the weapon at the target, close your eyes, move the weapon back and forth slightly, and assume the relaxed position again. Now open your eyes to see where the weapon is pointed. If the weapon is pointed to the left of the target, then you will need to adjust your body position slightly to the right, readjust, aim, close your eyes, move the weapon slightly, left and right, now open your eyes and see if you are in a more natural point of aim. This sounds time consuming; and it is. Once you have practiced this a few times, it will take no time at all. Lastly, once you become comfortable with getting into the position and develop a reference for your natural point of aim, you won't need to go through the entire drill. You will simply drop into a position that is close enough for government work. However, when trying to develop new positions, or test a rifle's ability to group tightly, I still follow the Natural Point of Aim procedure.

Once you have a basic understanding of natural point of aim, it is time to study the 4 Fundamentals of Marksmanship, the first of which is Position.

Position

Position, this important aspect of Marksmanship will be covered over and over. Be aggressive, drive the gun, and control the movement from target to target. Never let the weapon float and always remain in control.

For basic marksmanship, we will build a position that is durable. This means being able to stay in the position, something that does matter. If the position is hurting you, there will be unwanted pressure put on the weapon as well as the shooter may feel forced to break the shot before all aspects of the marksmanship procedure are met. Getting comfortable does not mean letting the rifle take over. You must stay in control of the weapon and in control of the firing sequence.

We also must ensure that we build a position that increases our chances of hitting the target but does not hinder the function of the rifle.

Figure 6.1B Don't grab the front of the magazine well.

Figure 6.1C Enhance the basic prone position by allowing the magazine to touch the inside of your forearm.

Figure 6.2A Modified prone position, note magazine on the ground, body flat as well.

Figure 6.1A Basic, prone position

Figure 6.2B Feet laying flat to limit disturbances to sight picture as you are shooting. Also a lower profile.

Sight Alignment

Sight Alignment, or properly aligning your front sight with your rear sight and then superimposing them on the target sounds simple enough, until we add the speed. It is not as important in combat shooting to always focus on the front sight. We must allow our eyes to bounce back and forth from sights to target. Unlike the bad guys, we have to target discriminate. This translates into determining who needs to hold one of your 5.56 bullets and who is just in the wrong place at the right time. If you focus too hard on your front sight during tactical scenarios, you will develop tunnel vision, giving up your ability to quickly identify a threat in your periphery.

Natural point of aim will also aid in making sights appear on the target rapidly. This is not point shooting. We quickly put our eyes on the target, drive the weapon to become aligned with our eyes, and watch the sights appear where we want them to. Once the sights are where we want, we move to the third fundamental of marksmanship.

Trigger Control

Proper Trigger Control is maintaining sufficient sight alignment while squeezing the trigger. Why sufficient sight alignment, not perfect sight alignment? There is a definitive difference in target engagements at 200 meters and those at 20 feet. During training we must determine what is the acceptable sight picture or flash sight picture for that particular distance. If you are using a red dot scope or conventional reticle tube scope, you will or should know exactly where your bullet is going to strike. If you are using open iron sights, then you may need to experiment on the flat range to determine what is acceptable. Can you get away with a rough sight picture looking over the peep rear sight at the front sight? Will you shoot high? How high?

One open sight technique for Close Quarters shooting is to flip your rear sight half-way between apertures. This will look weird, but basically you will be using the wings of the carrying handle like a crude rear pistol sight. Place the front sight in the huge notch and let'er fly. See Figure 6.3.

Figure 6.3 CQB Sighting with iron sights

Figure 6.4 Gripping the pistol grip as high as you can.

It is important to place the sights where you want to aim, not where you want to hit. Is this confusing? Your iron sights are 2.5 inches above your bore axis, or bore line, which requires adjusting your point of aim to compensate. This sight adjustment is called "hold off" or "offset." This equates to hitting low at contact distances. When making torso shots it may not be an important issue; however, what if you have a head shot with a no shoot (hostage) blocking a significant amount of the bad guys vitals? Now is not the time to figure out Hold Offs! Get on the range and get comfortable with your hold off's. If you plan to mount your scope on the carry handle of your carbine, then you will have even a greater degree of hold off. My suggestion is that if it is at all feasible, lower the scope. If your department won't allow this, where do you go? That's right, back to the range to practice even more.

For More information on Hold Off's and Trajectory see Chapter 8 Basic Ballistics.

Breathing

Hand in hand with trigger control is breathing. If you are not able to slightly control your breathing as you shoot, it can cause serious marksmanship errors. Some say you do not need to apply breathing to combat shooting; I am in total disagreement. Combat will require you to get control of your breathing more than any other type of shooting. During combat engagements you will have to gain control of not only your emotions, but also your breathing. I often catch students forgetting to breathe as they shoot a scenario. At the end of the engagement they are turning red or purple as their body is starved for O^2. This isn't good at all for the Combat Marksman. Your eyes are providing you most of your feedback as you shoot. Shooting is extremely visual and depriving your body of oxygen will almost immediately affect your eyes. So the most important body part pertaining to shooting is the first to go. This is not good; not good at all.

As you move into your shooting position try to steady your breathing. If the movement to your position has caused physical exertion and you are slightly winded, you must use some restraint to make yourself take a deep, cleansing, breathe before you start the trigger squeeze.

Let me walk you through the process. Build your position and align your sights. Once you have sight alignment it is time to start your trigger pull. Before this step, take a deep cleansing breath and blow it out. Next, take another breath and let your body slowly release this breath. Once you hit a sweet spot in the breath or the point where you feel all movement has stopped you want to momentarily stop breathing. This is the natural respiratory pause. This is not forced and it should be somewhat natural. At that point in the firing sequence you will need to let everything happen in a relatively short amount of time. If not, you will once again start to deprive your body of air, and shaking is soon to follow. Do not rush. If the shot will not break during the natural respiratory pause, then do it again. You will get better and better at making things happen at a slightly increased pace.

Follow Through

Trigger control and follow through are intertwined when applying the fundamentals of marksmanship. If you are jerking the trigger, it is very difficult to follow through for your next shot. By follow through we mean watching your sights throughout the firing sequence or at least being aware of where they are moving. If you are squeezing the trigger slowly straight to the rear of the receiver but as soon as the weapon fires, you quickly release the trigger to get ready to shoot again, then you are not following through. While practicing your trigger control and follow through it should go something like this: Align your sights, start your trigger squeeze while continuing to monitor where your sights are on the target. Once the weapon fires, be continuously aware of the sight to target relationship. Slowly release your trigger until you feel your disconnector click. Once the disconnector clicks, start your squeeze for the next shot. There is no need to completely take your finger off the trigger unless you are finished engaging that particular target.

Unlike other rifles, it doesn't hurt to get a little more finger on the trigger with the AR rifles. It helps significantly with heavy factory triggers, as well as with three round burst rifles, even when they are fired in semi auto mode.

Reading Your Sights

Reading your sights- always knowing where your sights are as the weapon fires. This requires a lot of training. This skill will increase your shooting speed significantly, particularly with distant targets (over 50 meters). Align your sights, squeeze the trigger, read your sights, shot breaks off the target, follow through and immediately re-engage. You save time since you don't have to wait for a reaction from the target. You already knew it was bad; now re-engage to remedy the situation. Is this creating bad habits by second guessing yourself? Definitely not; this will prove itself time and time again in training and, more importantly, you will be ready to engage a pop up shoot back target rapidly, making up important shots without hesitation.

Calling your shots- If you have difficulty knowing where your shots went, then you will need to be diligent in practicing calling your shots during your range sessions. This is not easy to learn; however, the more you shoot, the better you will get at calling where the bullets went. This is not only important at extended distances when engaging from the prone position, but it is also important when engaging at CQB distances.

If you have difficulty knowing where the shot went, then use a spotting scope to determine if your call was correct. Or, an easier method is to shoot with a friend who listens to your call and then tells you where the round actually hit. This may seem a bit time consuming, but all good tactical shooters are able to reliably call their shots.

7. GETTING ON PAPER
Getting Started

Once you have put together the correct AR package for your particular application, chosen your tactical gear, selected your ammunition, and studied the principles of Marksmanship, it is time to head to the range.

Target Selection

When selecting a target setup, you must be able to easily distinguish your hits. Also, make sure you take enough fresh targets to be able to replace the targets often. This will help you to constantly analyze your shooting without confusion. If you do not have additional targets, at least use tape or mark your bullet holes. This will save a significant amount of time if you spend any amount of time on the range. I also try to set the targets slightly lower on the stand since I plan to zero from the prone. Bottom line is: make sure your set up is safe and user friendly.

Do not use too large of a target. If the target is too large, then you will not have a repeatable hold for accuracy. You may want to put a small piece of tape in the center of the bullseye or use a small piece of tape on a white background. Find a method that allows you to hold at the same point repeatedly. This is of utmost importance during the zeroing process.

Check your Sights

You must also insure you have your sights zeroed or in the correct position for the range you are shooting. This is especially important if you are using an open sight with more than one aperture choice. Make sure the sight is in the correct position. As I stated in Chapter 3, these aperture centers may be different heights, so take this into consideration.

Along the same lines, always check that all optics and sighting systems are secure. If they are not secure, you will have a long day at the range; a long, bad, day. Don't be afraid to use Loc-tite thread locker to secure scope ring screws. You should also be using a torque wrench for attaching the rings to the rail system on your rifle. Do not use a mount system with thumb screws for security. To date, I have never observed a thumb screw mounting system that would reliably secure your scope. If you need a quick on/off system, use one of several available quick detach or throw lever mounting systems available from GG and G or LaRue Tactical.

Start Close

When you are finally ready to start your shooting, you should start at a relatively close range. I normally start at 25 meters to insure that I can get a decent group to center on the target before I head back to longer ranges. If you start at extended ranges, it may take you a while to get on paper; time is valuable, don't waste it.

I start with the prone position to eliminate other elements that may cause deviation in my zero. Prone will allow you to shoot tight groups, quickly get on paper, and then move back to extended ranges.

Once in position, fire three to five rounds to prove you are shooting somewhat of a group. At this range you should have one ragged hole; if not, continue to apply the fundamentals that we addressed in the previous chapter.

After firing three to five rounds, safe your weapon system and move down range to analyze your target. What is the outcome? What is the distance you need to move your sights to attain a correct zero? Do you want to zero at this yard line? Do you want to zero slightly low or high? For now, your focus should be on a tight group; it will be easy to move the zero to the correct location if you are able to print a good shot group.

Adjusting Sights

If you are shooting low or high, your front sight will be used to make this correction with Iron Sights. If you have a scope or optical sight, obviously all adjustments will be made at the scope.

Before making adjustments to your sighting system, have the right tool for the job. Sight changing tools for dialing your front sight as well as A1 rear sight are available. You should be able to pick one up relatively cheap at a gun show or local gun shop. If you are adjusting your scope, try to use the right tool to eliminate destroying your scope. If you ding your scope turret threads, you may not be able to put the protective cover back in place; this may lead to environmental issues with your system.

To move your group with the front sight, always remember to move the front sight the opposite direction that you would like your bullet impact to move. In other words, if you shoot low, your front sight needs to be lowered to correct the problem. If you shoot high, raise your front sight to move in the correct direction.

Adjusting your rear sight is just the opposite. Move the sight in the direction in which you would like your bullet to move. If you shoot low, raising your rear sight will send you in the right direction. There is no easy answer as to how much to move your sights. Iron sights as well as scopes and optics have several different ranges of adjustment.

If your Iron Sights have marked graduations for differentiating between ranges, ensure these markings coincide with your rifle/ammunition selection. Most tactical iron sight adjustments are rather crude; take this into account when attempting to zero. If you have purchased match sights, this will not be an issue, although match sights are definitely not a necessity for tactical rifle use.

Shooting Groups

Once you start the zeroing process, do not adjust your sights unless you are shooting consistent groups. Once you have the ability to tighten your groups, it is time to move on to getting the group exactly where you want it. If your bullets are not landing where you intend, continue to hold on the center of the bull or your aiming point. If you do not continue to hold on the same point, you will always be chasing your groups. Ideally you need to shoot extremely small groups at 25M if you are expecting to be confident at extended ranges.

Over Adjusting or Chasing Zeros

Once you have determined which direction you wish to move your group and it is, in fact, a group, it is time to move your sights. Now you must insure you know which direction to turn knobs or screws to get the desired results. If you are not sure, you should be able to check your owner's manual or check the web for information pertaining to your particular set up. If neither of these options is feasible you may need to experiment a bit. Once you have determined which is which, use a paint pen to mark your sights for ease of adjustment in the future. See Figure 7.1

When adjusting your sights, do not be over zealous. More times than not if I decide to make a large adjustment to get me exactly where I need to be, I will over adjust. This will put my group out

Figure 7.1 Paint Pen Markings

the other side of the bulls-eye or in some cases off target. This doesn't seem possible until you understand how some sighting systems work. They may be precise optical instruments or they may be crudely manufactured, yet highly reliable red dot systems. You may also notice unexpected zero shifts after firing just a few rounds, once again this could be your sighting system. Never adjust your scope without confirming the outcome. If you have shot your last group, yet decide to make a slight adjustment for some reason, then just to be sure, go shoot another group. It may come back to bite you in the backside. For competition this is fine, for man hunting, which may involve shooting near your mates or in the close vicinity of innocents, it is utterly unacceptable. Always shoot for confirmation. Never have a doubt or second thought when engaging in combat marksmanship.

Duty Ammunition

When you are zeroing, it is always best to zero with your duty ammunition. If you decide to shoot several different ammunition selections, it will only confuse the process. If you cannot afford enough duty ammo to attain all of your zeros, you may want to look at a cheaper alternative. Is your life worth it? Better yet, is your buddy's life worth it? I know what my choice would be.

Don't confuse the zero process with practice, if you decide to zero with duty ammo and practice with a cheap alternative, that is fine. However, remember this and do not tweak zeros during practice. Most ammunition will shoot closely enough to the same zero for most range work. In other words, unless you are shooting from long range, the cheaper ammo should suffice for your accuracy requirement during training.

Applying the Fundamentals

When you head to the range, it is the perfect time to apply the Fundamentals of Marksmanship. So, as you build your shooting position at 25 yards or meters, or any other distance, check your natural point of aim. Once the natural point of aim is on track and you have a solid position, double check sight alignment. Is your front sight crisp with the rear sight slightly fuzzy? Is the target fuzzy? The general rule is crisp front sight, fuzzy target. Maintain this rule until you move onto advanced shooting techniques.

Figure 7.2 Front sight crisp, target fuzzy

When your sights have been properly aligned and your breathing is correct, start your trigger squeeze. Upon the shot breaking, follow through, call your shot, and lastly, check your downrange performance. The fundamentals are very important during the zero phase. If you are using poor technique during zeroing, it will only get worse as you increase the distance and difficulty.

Positions

When zeroing, always try to use different positions that you will encounter in a tactical setting. I prefer to zero from the ground, shooting with my magazines on the ground. This is how I will shoot in a tactical engagement if possible. I also try to avoid shooting from a bench or while laying my rifle on concrete. These two positions, in particular, cause severe disruption in the natural recoil and accuracy of the weapon. If you shoot from a bench frequently, you will never notice a problem; however, once you apply your techniques to field or combat shooting, it may cause issues.

Analyze Your Target

After engaging your target, it is time to analyze the downrange performance. Not only do you need to look where your groups are forming, but you must also look at what they are doing. Are the shots stringing from top to bottom in an elongated group? If so, your breathing may be disrupting the sight picture or you may not be shooting during the natural respiratory pause. Are you occasionally throwing a shot wider than the rest of your group? Could this be trigger jerk, anticipating recoil, or tensing your shoulder as you shoot? Are you continuing to hold the trigger to the rear as you follow through? Are you quickly releasing the trigger? If so, this will make follow through difficult. Squeeze the trigger, let the weapon recoil, next slowly release the trigger until you hear or feel the disconnector click. This will not only help during the zeroing process; it will also help when shooting from extended ranges. Follow through, follow through, follow through. Minute changes can cause severe effects down range; this is why you need to apply the fundamentals every chance you get. It will only serve to make you a better overall Tactical Marksman.

Changing Zeros

It seems every natural or supernatural occurrence in our society affects your zero. This may seem irrational to some, but it is sometimes true.

Lighting affects zero, the old saying in the shooting world is, "lights up, sights up." This implies as the sun goes higher in the sky you may need to adjust your sights up, this pertains to the rear sight. So going one further step, if your group drops during a high noon range session don't be alarmed, this is a natural environmental effect. Other environmental effects that may cause a shift in your original zero are elevation, humidity, temperature, and of course, lighting conditions.

Ammunition can also cause changes. Rough handling can result in severe zero shift. Even dirty weapon systems can cause inexplicable deviation. What about too clean? Yes, it is true; a clean weapon may cause the first round or two to go unexpectedly in the wrong direction. With range time and copious notes you should be able to determine the root of the problem. Realistically, it is only a problem if you do not know how to compensate for the issue. In the sniper community this is the reason behind cold bore shooting. How and where will your rifle shoot when clean and cold?

If you have no plans on stretching your abilities to their maximum potential then these issues may never surface. I guarantee that if you shoot as much as you can, you will be confident when the day comes, no matter the conditions or distance. Know your zeros, and know your weapon system, inside and out.

8. BASIC BALLISTICS

Sight Height— How high your sights are

leaves the barrel it starts to feel the effects of gravity. It may appear that your bullet rises, but this illusion is caused by your line of sight to the target, the position of the barrel in the receiver with a slight upward angle, and the fact that your sights are looking down at the target. If this is confusing, imagine your bullet dropping as soon as it leaves the barrel. You may also need to picture your line of sight, a straight line, bisecting the bore axis, which is also a straight line. But do not confuse this with trajectory.

Trajectory

If this is confusing, picture a water hose held level to the ground. As the water leaves the hose, it is immediately affected by gravity; this is somewhat like your trajectory arc. To hit a distant water target, you will need to elevate or cant the hose upward. This causes the stream of water to exit the hose skyward and descend on the target in an arc. To hit a distant target with your rifle you will need to establish a similar sight picture.

The good news is that if you do a good job of zeroing your rifle for your shooting style and the situations that you encounter, then the issue of whether a bullet rises or falls is moot most of the time.

Since this is such an important part of achieving accuracy, I want to discuss rifle zeroing next.

Zeroing Your Rifle (Sighting In)

Whatever sights you use and whatever your habitual shooting distances, you must understand which path your bullet will follow. You must practice this while zeroing and then continue to hone these skills every time you visit the range. You must understand not only if you will hit a target at a specified range, but where you will hit that target. This is of utmost importance to attain accuracy for combat shooting. Hitting a silhouette at 300M should not be acceptable. You need to be able to hit smaller, more realistic targets.

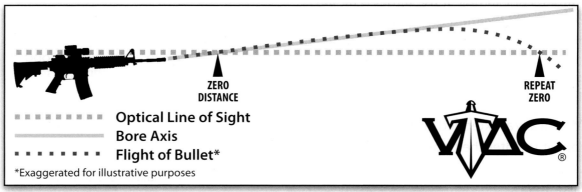

Figure 8.1 Bullet Arc Illustration Drawing

With a properly zeroed rifle the trajectory line, the bullet's path, will cross your line of sight at a particular range. For example, if you want your rifle zeroed at 25 meters, your sight will be held at the exact point you want your bullet to strike at 25 meters. See Figure 8.4 If you follow the trajectory for a 25M zero, you will see where your bullet path will be at various distances. Now, here's another reason for studying ballistics. If you know your bullet weight, ballistic coefficient, and muzzle velocity, then you will be able to construct a table that shows how far off target your bullet will strike at various distances. Zeroed properly, and with an appropriate cartridge for your shooting needs, your rifle can be very accurate at many distances with a single zero setting. I'll explain that further later in this chapter.

First, I want to discuss the basics of zeroing and introduce ballistics study. I'll have more to say in later chapters, but this will get you started.

Zeroing Basics

We discussed our zeroing procedures in Chapter 7, but here is a quick refresher. Zero from a likely shooting position, prone for example. Also maintain your basic fundamentals of marksmanship as you apply the zeroing process. Don't adjust your hold for zeroing. Maintain the correct, solid hold and move your sights as your group indicates you should. Once you have a group falling in the area of the target you desire to hit you should have a basic zero for one, pre designated distance. This is just the beginning for the Tactical Shooter. We don't live in a KD (known distance) world and you will never know at what distance your enemy target will appear. The basic zero is just a starting point.

25 Meter Zero

Some shooters decide on a 25 meter zero because it is easy. But what are the shortcomings of a 25 meter zero? One shortcoming is where do you hold beyond 25 meters? If your sight height is higher than normal – for instance you have a Trijicon 4 power ACOG mounted on a carry handle – your sights will be raised roughly 3.62 inches above the bore. With a 64 Grain bullet, traveling 2500 Fps, having a BC of .233, you will hit exactly where you aim at 25 Meters; however, at 100 Meters you will hit over 8 inches high, and at 200 Meters 11 inches high. Obviously this zero may not be the answer for the Combat Marksman.

This could cause serious confusion in a tactical scenario. However, with a 200 Meter zero on the same rifle, your trajectory will allow you to be dead on at 200 meters, 1.3 inches low at 25 meters, 2.6 inches high at 100 meters, leaving you with a lot less guessing.

Ideally, you would want to lower your sights closer to the bore to eliminate this problem or at least to alleviate the extremes. With the same trajectory parameters and a sight height of 2.5 inches (the height of your iron sights on the AR), your 200 Meter Zero would cause the bullet to hit 0.38 inches low at 25 Meters, 1.3 inches high at 50 Meters, 3.21 inches high at 100 Meters, 6 inches low at 250 Meters, and a little over 15 inches low at 300 Meters.

Following are several ballistic charts to get you started. Before you can know exactly where your rounds will hit at different meter lines, you will need to chronograph your ammo from your rifle. If this is not possible, don't be upset it your calculations are not exact. Every rifle will shoot the ammo at slightly different velocities. If a chronograph isn't available, a good starting point is to contact your Ammunition Manufacturer. They will be happy to assist you with basic information but, as stated earlier, it will not be exact in relationship to your rifle.

Figure 8.4 shows a Ballistic Chart for M855 Green Tip ammunition. This ammunition is issued to most Units in the United States military. Contrary to some gun magazines, M855 is a great ammunition. Not only is it loaded to military specifications, but it also works well in most situations soldiers find themselves in.

Figure 8.5 is a Chart for 77 Grain Sierra Match Kings, loaded by Black Hills Ammunition, and commonly referred to in the military as MK 262. This is another mil-spec ammunition that is used by members of the U.S. Special Operations.

Figure 8.6 has Ballistic information for Federal 62 Grain Tactical Bonded. This bullet and load are used extensively by government agencies.

Figure 8.8 is a chart that was made for a Police Department in California. As you can see, the chart is in yards since their normal range has markings in yards and not meters. This chart also has the actual information listed that was input into the ballistics program to attain the chart. Numerous conditions were entered to make the chart as accurate as possible. First velocity; this velocity was attained by firing several rifles with 10 rounds each to get the average for their issue rifles. Next is humidity for their area; average barometric pressure in inches of mercury, altitude, and lastly temperature. This chart will be relatively accurate year around for this department since they don't have huge weather changes. If you don't live in a California climate, then you may have to have several charts depending on the time of year. This is obviously important to soldiers who may zero their rifles in balmy Fort Bragg, NC, then head to the mountains of Afghanistan.

Figure 8.4 Ballistic Chart M855 Green Tip 62 Grain

Wind and Drop Chart

Load Information			
M855 Green Tip 62 Grain	Velocity: 2900 FPS		BC= 0.296
Sight Height 2.5"			

Range (Yards)	Rifle Zero				Wind Value				Range (Yards)
	25M Zero	50M Zero	100M Zero	200M Zero	5MPH	10MPH	15MPH	20MPH	
25	0	-1.09	-1.36	-0.89	0.04	0.07	0.11	0.14	25
50	2.17	0	-0.56	0.39	0.15	0.29	0.44	0.58	50
75	4	0.74	-0.09	1.32	0.33	0.67	1	1.33	75
100	5.46	1.11	0	1.89	0.6	1.2	1.8	2.4	100
150	7.18	0.66	-1.01	1.83	1.39	2.78	4.17	5.56	150
200	7.13	-1.56	-3.78	0	2.55	5.1	7.64	10.19	200
250	5.09	-5.78	-8.55	-3.83	4.11	8.21	12.32	16.42	250
300	0.77	-12.27	-15.6	-9.93	6.1	12.21	18.31	24.41	300
350	-6.15	-21.36	-25.25	-18.63	8.58	17.17	25.75	34.34	350
400	-16.03	-33.42	-37.86	-30.3	11.59	23.18	34.78	46.37	400
Zero Distance	25 M	50 M	100 M	200 M					
Repeat Zero	300 M	160 M	100 M	50 M					

NOTE: Distance drops and wind effects noted in inches.

By Kyle Lamb, © 2007 Viking Tactics, Inc.

Figure 8.5 Ballistic Chart 77 Grain Sierra Match King MK 262

Wind and Drop Chart

Load Information		
77 Grain Sierra Match King	Velocity: 2650 FPS	BC= 0.372
Sight Height 2.5"		

Range (Yards)	Rifle Zero				Wind Value				Range (Yards)
	25M Zero	50M Zero	100M Zero	200M Zero	5MPH	10MPH	15MPH	20MPH	
25	0	-1.06	-1.27	-0.68	0.03	0.07	0.1	0.13	25
50	2.11	0	-0.43	0.75	0.13	0.27	0.4	0.54	50
75	3.82	0.65	0	1.78	0.31	0.61	0.92	1.23	75
100	5.09	0.87	0	2.37	0.55	1.11	1.66	2.21	100
150	6.26	-0.07	-1.37	2.19	1.28	2.55	3.83	5.1	150
200	5.44	-3.01	-4.74	0	2.33	4.65	6.98	9.31	200
250	2.39	-8.17	-10.34	-4.41	3.73	7.46	11.2	14.93	250
300	-3.13	-15.8	-18.4	-11.29	5.52	11.04	16.55	22.07	300
350	-11.41	-26.19	-29.23	-20.92	7.72	15.43	23.15	30.86	350
400	-22.77	-39.67	-43.14	-33.65	10.36	20.71	31.07	41.42	400
Zero Distance	25 M	50 M	100 M	200 M					
Repeat Zero	275 M	150 M	75 M	35 M					

NOTE: Distance drops and wind effects noted in inches.

By Kyle Lamb, © 2007 Viking Tactics, Inc.

Figure 8.6 Ballistic Chart Federal 62 Grain Tactical Bonded

Wind and Drop Chart

Load Information		
Fed 62 Grain Bonded Tactical	Velocity: 2640 FPS	BC= 0.241
Sight Height 2.5"		

Range (Yards)	Rifle Zero				Wind Value				Range (Yards)
	25M Zero	50M Zero	100M Zero	200M Zero	5MPH	10MPH	15MPH	20MPH	
25	0	-1.05	-1.24	-0.54	0.05	0.1	0.15	0.2	25
50	2.1	0	-0.38	1.03	0.21	0.41	0.62	0.83	50
75	3.76	0.62	0.04	2.16	0.47	0.95	1.42	1.89	75
100	4.96	0.77	0	2.82	0.86	1.71	2.57	3.43	100
150	5.81	-0.49	-1.64	2.59	2	4.01	6.01	8.01	150
200	4.29	-4.11	-5.64	0	3.71	7.41	11.12	14.82	200
250	-0.02	-10.51	-12.43	-5.38	6.03	12.06	18.09	24.12	250
300	-7.63	-20.23	-22.53	-14.07	9.05	18.09	27.14	36.19	300
350	-19.17	-33.86	-36.54	-26.67	12.83	25.66	38.48	51.31	350
400	-35.35	-52.14	-55.21	-43.93	17.44	34.87	52.31	69.75	400
Zero Distance	25 M	50 M	100 M	200 M					
Repeat Zero	250 M	140 M	70 M	33 M					

NOTE: Distance drops and wind effects noted in inches.

By Kyle Lamb, © 2007 Viking Tactics, Inc.

Figure 8.8 Ballistic Chart XM 193 55 Grain

XXXXX Police Department
Wind and Drop Chart

Load Information		
Federal XM193 55 Grain	Velocity: 2940 FPS	BC= 0.338
Humidity: 67%	Baro Pressure: 29.98 in/HG	Sight Height 2.5"
Altitude: 42 Feet	Temperature: 64 Degrees F	

Range (Yards)	Rifle Zero				Wind Value				Range (Yards)
	25YD Zero	50YD Zero	100YD Zero	200YD Zero	5MPH	10MPH	15MPH	20MPH	
25	0	-1.12	-1.47	-1.16	0.03	0.06	0.08	0.11	25
50	2.24	0	-0.7	-0.09	0.11	0.22	0.33	0.45	50
75	4.2	0.84	-0.2	0.7	0.25	0.51	0.76	1.01	75
100	5.86	1.39	0	1.21	0.46	0.91	1.37	1.82	100
150	8.26	1.56	-0.53	1.28	1.05	2.1	3.15	4.2	150
200	9.31	0.36	-2.42	0	1.91	3.82	5.73	7.65	200
250	8.85	-2.33	-5.81	-2.79	3.06	6.12	9.19	12.25	250
300	6.71	-6.71	-10.88	-7.25	4.53	9.05	13.58	18.1	300
350	2.7	-12.95	-17.82	-13.58	6.33	12.65	18.98	25.31	350
400	-3.39	-21.28	-26.84	-22.01	8.49	16.99	25.48	33.98	400

Zero Distance	25 Yards	50 Yards	100 Yards	200 Yards
Repeat Zero	374 Yards	210 Yards	100 Yards	53 Yards

NOTE: Distance drops and wind effects noted in inches.

By Kyle Lamb, © 2005 Viking Tactics, Inc.

The Best Zero

Often I hear, "Which zero do you recommend?" This is a simple question. I recommend a zero for which you know all of your hold offs. If you require a slide rule to figure out where to hold, then that is not a good zero. I would prefer to shoot a 200 meter zero on most occasions. The key here is most. There are times when a 200 meter zero can be detrimental to your performance. An example would be while shooting specialty ammunitions, such as subsonic. If the ammo is not capable of performing at that distance, change zeros.

The 200 meter zero generally allows you to easily identify your holds without taking your sights off the target. Here is an example, shooting M855 62 grain ammunition at 2900 fps. If I use a 100 Meter zero, I would need to hold over 15 inches high at 300 yards to hit the target. Comparatively, with a 200 Meter zero, I need only hold 10 inches high to obtain a hit. Five inches may not sound like a lot of difference, but it is. Depending on which sighting system you are using, the target may be completely obscured to achieve a hit. This isn't a good thing in a tactical confrontation.

If zeroing at 200 meters isn't practical for you when you set up your rifle, 50 meters is a pretty good compromise. With the same M855 ammunition discussed above, a 50 meter zero is pretty good out to 200 meters. At 100 meters you would be just over one inch high and at 200 meters your hold off would be about 1.5 inches – not too hard to manage. By 300 meters, however, this compromise zero would require more than a foot of hold off to hit the target, a little too far on the high end for me.

In general you should select a zero for best flexibility and versatility, not for one particular factor. I prefer a zero that doesn't require math to attain a quick hit, at least out to 300 meters. If your plan is to always shoot from a distance of 50 meters or less, it may be time for a dose of reality. Just when you least expect him, there he is, standing right beside our good friend, Mr. Murphy. And by the way, he is standing at 200 meters.

Ballistic Software

When attempting to attain accurate ballistic data for your rifle, you will need a software program to aid in this pursuit. In the old days we simply went to the range and shot groups at all different meter lines until we thought we knew what to expect from our particular loading. Those days are in the past for me. I have found that using one of the ballistic calculators, PDA's (palm pilot), or other software, will help to decrease the amount of time spent chasing, and confirming zeros. Several items are available on the market to help you successfully determine what will happen once your bullet leaves the barrel.

After spending some time at the range establishing velocity baselines from my particular weapon, with several loads, I will head to the Ranch to determine a starting point. I find that a little pre-planning before each range session allows me to be more efficient when I finally have time to spend shooting. Before heading back to the range, I will run bullet data through either my home computer or a small hand held Palm Pilot using the ATRAG or Sierra Infinity External Ballistic software.

This will help me to determine if any of the selected bullets will perform. The manufacturer's data may say one thing but I will determine if the bullet will, in fact, meet my needs with my current rifle setup.

If you do not input good data, you will not receive good output. Next thing you know you are right back where you started. If you decide to use manufacturer's velocities, or swag it, remember this when trying to engage targets at extended ranges successfully: it won't be as easy to figure out where to start.

Wind Effects

I often tell shooters that I like to shoot the heavier bullets to help with the wind, even though they don't perform a whole lot better than the lighter bullets. I guess I enjoy the peace of mind. If you were to shoot a 77 Grain Sierra at 2,650 Fps, and compare its wind drift to a 62 grain M855 bullet fired at 2900Fps, you would only decrease your hold off for a 10 mph full value cross wind by about 1 inch at 300 Meters. When comparing the same bullets and a 20mph wind, the difference at 300 Meters is less than 2 1/2 inches, with the 77 Grain bullet winning. This isn't a huge amount when you are talking about 300 Meters. As your distances increase, so will the help of the heavy bullets. You need to determine how often you will shoot at these extended ranges.

How Far is too far?

At what distance can you expect to reliably hit a man size target? That is the age old question that has helped to generate more than a handful of Urban Legends. I have seen very talented competition shooters hit 10 inch plates at 500 yards. I have also seen those same competition shooters have one heck of a time if the wind kicks up, or the mirage starts to decrease visibility.

I hate to be accused of nay saying; however, when a soldier says he was hitting skinny individuals at ranges past 500 meters but they just wouldn't go down due to the poor wounding characteristics of the M855 bullet I have to raise the BS flag. At 500 meters it is hard to hit an E-type silhouette, let alone a man moving across the desert. Not to mention what problems you are experiencing with the plunging of your bullet. This plunging fire increases your need for accurate range estimation to make a successful hit. But now we're getting into terminal ballistics, which I will cover later in the book.

9. STAND UP AND FIGHT LIKE A MAN!
M-4 / AR STANCE / GRIP

Ever since I met my first truly tactical, combat shooting instructor, I have been a huge fan of the isosceles shooting stance. This stance works extremely well when you are shooting a pistol since both hands need to be almost symmetrical. I guess this is why the OSS used this stance; they were truly pioneers in pistol craft. Too bad it has taken so many years to come full circle, or almost, for some of you!

Figure 9.1 Jedburghs shooting isosceles, courtesy US Army

Stance

Shooting a combat rifle on the other hand is a different story. When you square your body to the target while shooting a rifle, it seriously limits your ability to do several things. See Figure 9.2 First, you cannot quickly drive the gun from target to target to eliminate several threat targets. You also are not able to keep your support hand far enough out on the rifle. This not only impedes recoil control, but also weapon retention.

Figure 9.2 Square shooting stance

The best rule of thumb with the combat carbine or subgun is to get in a boxers stance. Now mount the rifle. When you have your support side foot closer to the target you will have a much more aggressive stance. Your fighting stance is your shooting stance and your shooting stance is your fighting stance.

Figure 9.3A Fighting stance without the rifle.

Figure 9.3B Fighting stance with the rifle.

Once you feel that you are in an aggressive stance, take a look at the rest of your body. With the aggressive stance you should have a lot of weight on the balls of your feet. You shouldn't have so much weight on the balls of your feet that you feel like you are about to fall forward, but you should have enough weight on the balls of your feet that you are not getting rocked back by each shot. Having weight already pre-positioned on the balls of your feet will allow you to immediately start foot movement without first having to shift weight forward and then step. It will also help you to stay very aggressive during movement. Your knees should also be slightly bent. They will act as shock absorbers while stationary and, even more, during movement.

Your head should stay as upright as possible. This action will aide you in focusing clearly by causing you to look straight forward form the eye socket, just the way the Man upstairs designed your eyes to work. An upright head also opens up your total field of view. This is extremely important during combat operations in order to prevent stress induced tunnel vision, something that will hinder your performance even more.

For most tactical shooters, conducting an operation or training for an operation usually involves the use of helmets, goggles, gas masks, or night vision devices. An upright head allows you to operate efficiently with all of your gear while shooting. While maintaining an upright head, also try to keep your nose near the charging handle.

Figure 9.4A Poor head position, neck is craned forward. This position will not work well when you are wearing a helmet or using Night Vision Goggles.

Figure 9.4B Good head position, this position will allow you to look through the center of your eyeballs, through the center of your goggles and have much better field of view.

Hand Positioning

Hang on to your rifle!! Not just for retention purposes. During close quarter engagements the rifle must be violently moved from target to target; it's important to have a good grip on your rifle. Remember smooth is fast, but slow is just slow. The shooting hand can also apply more pressure than with a pistol and get away with it. The AR variants allow you to get a little more hand on the grip than most weapons allow. You should grip the rifle as high as you can. See Figure 9.5 High grip position. By this, I mean that there should be no gap between your hand and the rear portion of the rifle's pistol grip. You also need to ensure there is no gap between the middle finger of the firing hand and the bottom of the trigger guard. This is also key during malfunction clearances or during any procedure that only allows the shooting hand to hold the weight of the rifle.

Figure 9.5 Gripping the pistol grip

The strong hand should also apply rearward pressure, pulling the weapon into your shoulder. As you pull to the rear, drop the strong side elbow. See Figure 9.6A and 9.6B Elbow up and elbow down. Shooting a M4 in anger is different than National Match target shooting; drop the elbow to make yourself a smaller target. Dropping the elbow will also keep you out of the way of your mates while helping to reduce the chance of continuously bumping your elbow against door frames and walls while moving.

Figure 9.6A Elbow up, giving the bad guy a target to shoot at.

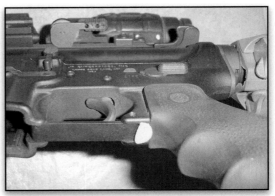

Figure 9.7 Plugging gap with ear plug

Figure 9.6B Elbow down, tight against your body. Works better in confined spaces.

If you begin gripping with the high grip as we have described, you may start to develop a sore middle finger. If this happens, you can buy a different pistol grip that covers that portion of the exposed trigger guard. For a quick, range fix, simply plug this space with a foam ear plug. See Figure 9.7

The support hand should be positioned far enough forward on the hand guards to aid in weapon retention as well as during target to target acquisition. If you use a forward pistol grip, slide it out, not so far that it is uncomfortable, but get it where it assists in driving the weapon. Do not grab the front of the magazine well; this is fine for one shot but is not a technique to be employed for combat operations. See Figure 9.8 Showing poor support hand hold of magazine. I have also seen this type of hand positioning induce malfunctions when the shooters fingers or thumb ends up covering the ejection port of the weapon. See Figure 9.10.

Figure 9.10 Fingers covering ejection port.

When you fight with a stick or staff, you spread your hands to give yourself strength as well as speed for hitting an adversary or quickly changing directions. See Figure 9.9A. The support hand should also be positioned slightly to the left from the bottom center of the hand guard if you are right handed. It should be placed slightly to the right of bottom, center if you are a south paw. Don't just lay the weapon across your hand; this is another National Match technique.

Figure 9.8 Grabbing the front of the magazine

Figure 9.9A Support hand moved forward to a fighting grip, with pistol forward pistol grip

Forward Pistol Grips

The use of the forward pistol grip is a touchy subject to some people. Will it make you shoot any better? More than likely not, it will, however, give the tactical shooter something to hang onto in adverse conditions. Climbing a ladder is a prime example of a time when it is nice to have a pistol grip on the tactical carbine. The forward pistol grip allows the tactical shooter to have a place to put his support hand when his hand guard starts to fill up with lights and lasers. This extra handle will aid in weapon retention for close quarters situations. They say pistol grips are for CDI (chicks dig it) points. However true that may be, we aren't trying to become the Prom King. As tactical shooters, we simply need a place to put our support hand. The forward pistol grip can be used to your advantage more times than not. I have even used it to hook the front of a pod (seat on a MH-6 Little Bird), so I didn't have to hang on as I was riding on the outside of a helicopter. I did this while adjusting my goggles; look Mom, no hands.

See Figure 9.11 Pistol grip types as well as different hand positioning.

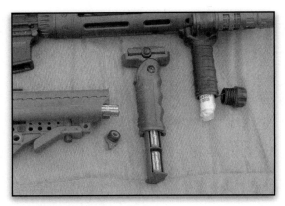

Figure 9.11 CADEX Folding Grip and Tango Down Pistol Grip

Figure 9.12A Alternate hand positioning on the pistol grip.

Figure 9.12B Gripping the pistol grip to activate light.

Figure 9.12C Opposite view of the conventional pistol grip.

Mounting the Weapon

Once you are comfortable with your stance and grip, it is time to aggressively mount the rifle. By mounting, I imply bringing the weapon from the low ready to your shoulder and having a correct sight picture. As you mount the rifle, bring the weapon to you. Don't mount the weapon and then drop your head into position. This is extremely slow and isn't as repeatable! Learn to position your head in such a manner that your rifle comes to you!

Correct Head Positioning

Figure 9.13A Head positioned in such a manner that the weapon will simply move to meet the head.

Figure 9.13B As you start to mount the weapon, continue to focus on the threat.

Figure 9.13E As the weapon comes up you are still looking over the sights.

Figure 9.13C Once the weapon has reached your cheek, you are ready to fire. Looking through your sights at the threat target.

Figure 9.13F Still unable to shoot since you can't see your sights.

Incorrect Head Positioning

Figure 9.13D Incorrect head positioning. Not aggressive or ready to drive the weapon to the threat.

Figure 9.13G Head is dropped into position to take the shot, but this technique is extremely slow and not as repeatable.

Left Hand Safety Manipulation

Once you are ready for the range, you must engage your safety every time you bring the weapon off target. It is very important to remember that the safety only comes off once you have acquired a target. By acquired, I specifically mean you have the target in your sights. If you are left handed, try disengaging your safety with your index finger knuckle rather then the thumb. Placing your left thumb on the left side of the weapon does not bode well for weapons retention and aggressive shooting. See Figure 9.14. Try your best to disengage with the left index knuckle. See Figure 9.15. There are also ambidextrous safeties available from several manufacturers; this will allow the left handed shooter to disengage and engage the safety.

Figure 9.15A Correct safety disengagement, knuckle of index finger positioned over safety.

Figure 9.15B As the target is acquired, the safety selector is pushed to the fire position. Note weapon can still be retained with complete grip.

Figure 9.14A Incorrect safety disengagement, disengaging the safety with the firing thumb.

Figure 9.15C Once your index finger has pushed the safety into the fire position, you will be in a position to fire the weapon.

Figure 9.14B Giving away a good weapons retention grip.

Muzzle Up or Down?

Does it really matter if we carry muzzle up or down? That is the question. The answer I often hear is, "whatever you are comfortable with." This may be true when you are hunting pheasants in South Dakota with your Pa's Ithaca Model 37, but we've come a long way baby!! In almost all tactical scenarios, and most definitely in real world situations, muzzle down is the ticket.

Let's walk through a MOUT scenario. How do we get there? For some lucky modern day knights, their trusty steed is a JP8 guzzling, rotary winged horse. You definitely need to be muzzle down inside of any helicopter. If your infil platform is a vehicle, car, or truck, muzzle down is still a good call. It allows you to safely get in and out of the vehicle without sweeping your team mates. A weapon with a muzzle down also tends to stay that way in a crash or ramming situation.

Once you get to your crisis site or target, it still pays to stay muzzle down. This will give you a lower profile as you move from position to position. If you do have to run, it is much easier to move fast with the muzzle down. And as I said before, there is also a better chance of not flagging your teammates again.

When moving in and around buildings, it begins to become more and more important to be ready, and ready is not with your muzzle pointed in the sky.

Weapons Retention

Primary weapon retention is always a hot topic. With the muzzle up, it is an inevitable struggle. With the muzzle down, you have options. If you are quick with a figure 8 motion this may surprise the bad guy into letting go. You may also choose to front kick this crazy man, or simply drop to a knee quickly to give you a better shot at his pelvis. You could also use the pull away technique. If he is trying to take my weapon, my guess is he doesn't just want to see what the serial number is.

Figure 9.16A Suspect grabs barrel.

Figure 9.16B Shooter prepares to pull weapon away.

Figure 9.16C Quick jerk to the rear will help to a line barrel with threat.

Figure 9.16F Drop as quickly as possible to a knee to align sights with threat

Figure 9.16D Pull down on the weapon slightly.

Figure 9.16G Engage threat as necessary.

Figure 9.16E Once you pull down, threat will pull up.

Figure 9.16H Threat grabs weapon.

Figure 9.16I Prepare to front kick.

Figure 9.16J Front kick, be careful to remain balanced.

Figure 9.16K Engage threat as necessary.

Lastly you may need to release your long gun with your strong hand and draw your pistol. At this point don't just give him something else to grab. Be smart.

While moving through a building there may be a few incidents where muzzle up makes sense. Be smart. If you or a teammate were to have an AD (accidental discharge) you at least will know where it (the bullet) went. Muzzle up- who knows! I know, I know, that would never happen to me; there are a lot of folks out there that said this in the past. Now they are sitting down to a fine piece of baked crow.

I often see medical personnel, as well as breachers, sling their weapons muzzle up. Their explanation is "I don't want my muzzle in the dirt". Oh sorry, just stick it up my nose or behind my ear as you work to patch up a teammate or hostage. If you are worried about the muzzle put a 25 cent muzzle cap on your weapon or a piece of 100mph tape (LAPES tape to you Air Force guys – duct tape to everyone else). See Figure 9.17. Even if an unprotected muzzle finds some mud, it will take an awful lot to plug your weapons barrel sufficiently enough to blow up the weapon; your flash hider might look funny, but the rifle normally will be ok.

Figure 9.17 Muzzle Cap

Carrying a rifle muzzle down also increases your field of view while looking for the threat or target indicators. Moving the muzzle from a down position is quicker and is more apt to allow a good stock to cheek weld.

10. WE DON'T LIVE IN A PRONE WORLD
Tactical Shooting Positions

In order to be a tactical marksman, you must have the ability to quickly provide accurate fire on a threat, no matter the position. If you feel it is not necessary to practice longer range engagements because you are on an entry team, you are sadly mistaken. It is a colossal advantage to be able to shoot from any position and at any range inside your weapons maximum effective range and still be confident in the outcome. With practice, you will be able to determine exactly what parameters must be met in order to put effective fire on the threat.

Basic position shooting is normally taught in shooting schools for long range weapons. Invariably, you will practice shooting from the prone position 90 percent of the time. But how often in a real scenario are you even able to get into this perfect prone position? During most carbine courses, the prone position is used to confirm or establish a weapon zero and then you never shoot from any position other than a standing offhand position.

If you are attempting to cover the whole spectrum of tactical shooting, then you must practice the most awkward and obscure positions you can think of. How do you decide which position you will need to learn? Look around you. Is there a car to shoot under, a corner to clear, a curb to provide some sort of cover, or is there just a pile of logs that happens to be offering you some protection? When you head to the range, have a list of items you want to shoot over, under, or around. Build a barricade that will help simulate most of the positions. With the help of my good friend, Bennie Cooley, we have built a barricade that covers most positions. This barricade is also made of wood, so that if you do happen to shoot the barricade, you won't be injured. (See VTAC barricade illustration and photos on the next page.)

As you can see, this barricade will aid in easily setting up your range to facilitate great barricaded, position shooting. Use your imagination to develop positions that you can use in real-world situations.

I will cover numerous positions, starting with prone and working all the way to off hand. These positions will not work with all equipment or body types. Try them, see what works, and modify the positions to fit your build, your equipment selection, and your particular scenario.

Natural Point of Aim

In Chapter 6, Fundamentals of Marksmanship, we covered Natural point of aim. Please review this information. Natural point of aim plays a significant role in quickly developing a shooting position in a fluid situation. In tactical scenarios, you must be able to quickly get rounds on target. It will slow you down if you have to jockey positions in order to establish a natural point of aim. Slowing

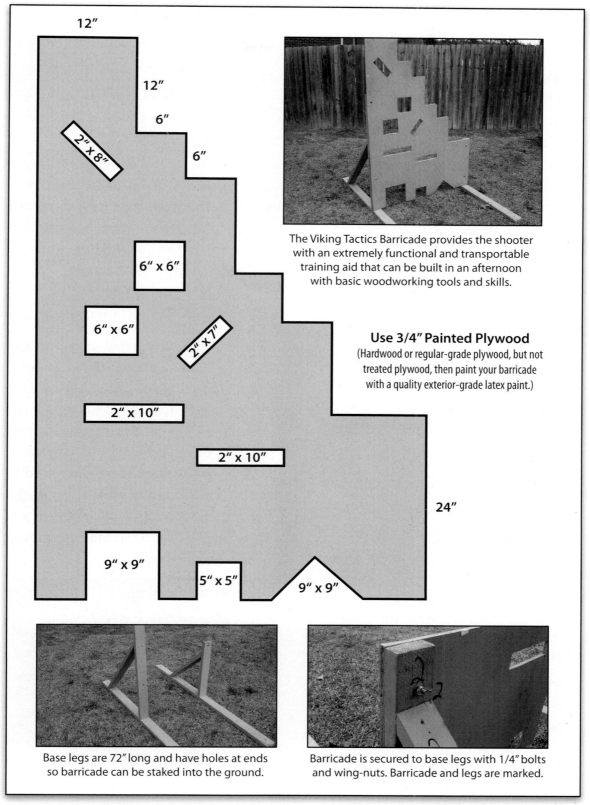

Figure 10.1 VTAC Barricade

down only gives the threat more time to seek cover, or worse yet, to engage you. When building several of these positions, you will notice that you may have to turn your body significantly more than in other positions to attain the desired, natural point of aim effect. Keep working until the position allows you to meet your goal. This goal should be to quickly engage several targets with accurate fire. Whether or not you have cover, speed is very important during these engagements.

> *Practicing positions gear.* When I initially start developing or practicing positions, I like to do it slick. Slick meaning without wearing my assault kit. Once the positions are coming together, I add more kit to see what slight changes I must make in order to get the position to work. Your position may not change a lot, but it will evolve slightly to deal with the additional gear. Ensure that you shoot with your helmet and communications gear in place as well. The helmet can be one of the biggest hindrances to your positions, so it is important that you practice to overcome this hurdle. Do you wear gloves for operations? If so, then you should work on the range with gloves as well. The addition of gear may also cause you fatigue, which will influence the stability and durability of these positions. It is better to train through these problems than to deal with them for the first time during an actual mission. Remember to always "train as you fight."

Prone

Going prone is a very natural shooting position. This position is used by almost everyone around the world in some form or another. What can we do to enhance this position with the AR variant?

Body Positioning

One of the first things that we can do is to align our body slightly more in line with our rifle than normal. This will allow us to drive the gun from target to target if the need arises. You need to let your body lay as flat as possible. Unless you have raised your heart rate, then there is no need to cock a leg forward to lift the diaphragm off of the ground. This will also help you if you need to quickly lift the rifle to move it onto another target.

Figure 10.2 Prone Position

Weapon Positioning

A combat prone position is nothing like a National Match type position. There are no rules that tell us what can and cannot touch the ground, so get the magazine on the ground if possible. This is taboo to most shooters, but the best tactical shooters in the world have reliable gear that also allows them to use the magazine as a rest. If your weapon malfunctions while resting the magazines on the deck, replace them. While in the prone, your sling should provide you with some support, and if it does not, tighten it so it does. The Tactical sling should be more than a carry strap.

Foot Position

Once you are as flat as possible, check your feet. Are they lying flat on the ground as well? If not, they should. As you fire your weapon the recoil will move your feet, translating into wasted time waiting for your foot tremor to stop. See Figure 10.3 Poor Foot Positioning. This may seem insignificant but every little bit of time we can gain may be the difference between shooting and being shot. I like the first option better.

Figure 10.3 Poor Foot Positioning.

Support Hand

Take a look at your support hand. Is it grasping the front of the magazine while prone? If so, slide it out on the hand guard. This will help significantly with recoil control. This technique will also aid the tactical shooter when tracking moving targets, or transitioning the weapon from one target to another. See Figure 10.4A-10.4B. This hand does not need to be relaxed unless you are shooting extreme distances. It also helps to slightly pull the weapon back into your shoulder. As you do this, you need to bear down with your cheek weld as well. When you tighten up, make sure your nose is on the charging handle if at all possible. This will also help with recoil control. Like I keep saying, every little bit helps.

Figure 10.4A Improper Support Hand Placement.

Figure 10.4B Proper Support Hand Placement.

ALTERNATE PRONE POSITIONS

High Prone

If you are unable to get into the combat prone position with your magazine on the ground, you may need to somewhat modify this position. See Figure 10.5 High Prone. If, for some reason, you need a little elevation, lift your body slightly up on your elbows. This high prone position is a lot more stable if you have a tight sling. You can also gain more stabilization by letting the magazine rest against the inner side of your support side forearm. If it feels like you are leaning to the support side, you may need to adjust your body position to shoot more in line with that support arm. See Figure 10.6A-10.6B. When shooting from these positions, always wear some sort of elbow pad to alleviate any busted elbows. It will make your shooting sessions much more enjoyable, leading to better learning.

Figure 10.5 High Prone

Figure 10.6A Magazine touching inside of forearm.

Figure 10.6B Magazine touching inside of forearm, front view.

Rollover Prone

Next up is a favorite position of mine, rollover prone. See Figure 10.7A-B Rollover Prone. This position will allow you to shoot under obstacles you never imagined you could shoot under. Start by laying down on your strong side with your elbow tucked under your body. See Figure 10.8A. Notice in Figure 10.8B, if you do not tuck your elbow you are not able to get as low on the rifle. Your left knee will be pulled slightly up, with your pelvis facing the ground. Rehearse this position with the gear that you will wear on a mission. This will insure you can build this position when needed. This will seem uncomfortable at first, but after a while, you will find it is second nature.

You won't be able to get a great stock to cheek weld, so this will feel unfamiliar at first. Practice will build confidence. It is also important to try to align your scope's red dot with the iron front sight. This will help to eliminate parallax. The support hand can provide support by grasping your forward pistol grip or using the hand guard almost like a bridge with a pool cue. See Figure 10.9A-D. Whatever works to hold the weapon on target is sufficient. When practicing this position, make sure you wear shooting glasses or goggles. If you have a compensator on your rifle, there will be a spray of rocks or dirt in your face. I recommend you use a flash hider rather than a muzzle brake. There will be less chance of obstructing your view with all the dirt or dust. Lastly, check the clearance of your ejection port. If you are too close to the ground you may induce a stoppage, a failure to eject, more commonly called a stovepipe.

Figure 10.7B Rollover Prone, notice pelvis is facing the ground.

Elbow Positioning

Figure 10.8A Elbow Tucked under body

Figure 10.7A Rollover Prone. Extremely low position.

Figure 10.8B Elbow not tucked, not able to get as low.

Support Hand During Rollover Prone

Figure 10.9A Support Hand during Rollover prone.

Figure 10.9B Alternate support hand position for rollover prone.

Figure 10.9C Sometimes just making a fist will help you to quickly get the sights on the target.

Figure 10.9D Note fingers being extended to work like a bipod.

If you are shooting from Rollover Prone at close range, 25 meters or closer, you won't have an issue with your hold off's. If you are shooting at more extended ranges, then you will need to confirm where your rifle will hit with the rifle laid 90 degrees to the side. You need to take into account that the bullets trajectory will be off considerably at distance. If you are rolling the rifle 90 degrees to the right, the bullet will strike low to the right. So, in order to hit the target where you intend, you will need to hold high left. An example of this situation would be as follows: If shooting at an 8 inch plate, you would need to hold at about 11 o'clock of that plate to hit in the center of the plate. This is contingent on your rifle's zero; but, it is a good rule of thumb. If you are rolling the rifle to the left 90 degrees, it will be just the opposite, or, in other words, you will need to hold at 1-2 o'clock on the edge of the plate. When referencing clock directions 12 o'clock is always straight up, no matter your body position.

Reverse Rollover Prone

If the situation does not permit you to get into a rollover prone, you may need to reverse the position and shoot reverse rollover prone. If, for example, you are trying to use a street curb for cover while your body is in the street. See Figure 10.11. You may need to shoot reverse rollover to maintain the proper use of cover. In order to accomplish this, you may need to lie down on your back with the rifle on your strong side shoulder. Your pelvis will now be facing up and you may also have issues getting a good cheek weld here. Test the position to see what idiosyncrasies you encounter. Is your gear set up to allow you to lie flat on your back? How low can you get your elbow? You must always look realistically at these positions. If your elbow is up, you are giving the bad guy a target. See Figure 10.12A-B. This position can also be used when shooting around a low wall where you feel it is easier to lay out on your back to shoot, rather than shoot from a conventional prone position.

Figure 10.12A Reverse rollover prone elbow up.

Figure 10.12B Reverse rollover prone elbow down.

Figure 10.11A Reverse rollover prone by curb.

Figure 10.11B Reverse rollover prone, note elbow and feet down behind cover. Lay your feet flat if you have to.

SBU Prone
(Prone with Weapon Canted)

I have heard this position called some really whacky names such as SBU, SBT, and Navy Prone. I don't care what it is called. All that I am interested in is how it will add to my tactical shooting position repertoire. This position resembles your regular or conventional prone. The only exception is you will cant your rifle towards you. For right handed shooters this means the top of the rifle is canted to the left. See Figure 10.13A-B and Figure 10.14. If possible, the magazine will be resting on the ground. You will normally not be able to have a good stock to check weld; but, if you properly align your sights, you will hit what you are aiming at. If you are wearing a helmet, you may have to push it back on your head a little to get an eye on your sights. You may also need to use your bicep on the butt of your weapon. Don't get excited if you can't get the stock to your shoulder. Butt stock on bicep for SBU Prone. This position will allow you a lower profile when cover is limited. It may also be your answer if you are unable to assume rollover prone. This position will not allow you to present your weapon as low; but, it will allow your body to give less of a silhouette for the bad guys to shoot at.

Figure 10.13B SBU Prone front view. Note buttstock on bicep, not shoulder.

See Figure 10.16 SBU Prone buttstock not touching bicep or shoulder.

Figure 10.14 Left Handed SBU Prone.

Figure 10.16 SBU Prone buttstock not touching bicep or shoulder.

Figure 10.13A Right Handed SBU Prone.

Figure 10.17 Close up of weapon laying on the right side, how low can we go with the ejection port.

Shooting under cars. There are a few things to be aware of when using any of these modified prone positions under vehicles to engage targets. First of all, if you can see the bad guy, then he can see you. This is an important aspect for acquiring these positions. You must be able to get into the position quickly. This reduces the time in which the target does not have a weapon trained on it.

As you move into these positions around vehicles, give yourself some standoff. Stay back from the side of the vehicle so that you have at least 2 feet between the end of your muzzle and the side of the car. If you have ever shot your rifle with the muzzle under the vehicle, then you will know what happens. The muzzle blast from the weapon knocks every bit of dirt loose from the last 10,000 miles. This dirt either drops on you or, at the very least, obscures your sights. Obviously, the obscured sights are of our primary concern. If you have properly distanced yourself from the side of the vehicle then this will not be an issue. Take your rifle to the range to try this and see what happens. Just make sure that you have a clear shot to the target without any portion of the vehicle between the muzzle and the target. I have seen many tires flattened as shooters experiment with these positions.

When you pick cars to play with, try to use the lowest vehicles you can find. You will be surprised how low you can actually get with your rifle.

Figure 10.18A Rollover prone under vehicle.

Figure 10.18B SBU Prone under car.

Figure 10.18C Broke Back Mountain under car.

Figure 10.18D Extremely low but possible.

Figure 10.18E Under vehicle view.

Rice Paddy Prone

Rice Paddy Prone- Judging by the name, this is a position that allows you to have a relatively stable platform, while not completely wallowing in a rice paddy. This position is also known as the Kimchi squat. If the situation dictates the need for a position that is easy to assume and allows quick movement once your shooting is complete, then this may be a position of interest. This is also an excellent position for hunting big game in a marshy environment. As with all other positions, these are tools for your tool box. Try to use as many positions as you can develop. You never know when one of these positions might come in handy. Rice paddy prone is a squatting position with your knees equally bent and triceps supporting the weight of your upper body on the knees. To be successful with this position, you have to be slightly flexible and have good muscle tone in the thighs. See Figure 10.20 Rice Paddy Prone. Once you have lowered into the position, you should not require a lot of muscle support to maintain it. Your legs should be about shoulder width apart with the toes pointed somewhat parallel. Keeping your feet flat will help with stability once the shot has been fired. You may need to drop your elbows inside your knees for additional rigidity. See Figure 10.21A-B. Once your elbows are in position, put slight pressure with the inside of your knees onto the outside of your triceps or elbows. You should feel the position tighten. One limiting factor with this position is the inability to acquire and engage targets in a wide arc. Your field of fire will be somewhat limited, but, as I said earlier, this is another good, versatile, tool for the toolbox. You will have to work on establishing where your natural point of aim is with this position. It is a deceiving position and you will more than likely have to adjust your body drastically the first few times you get into the position in order to successfully engage your target.

Figure 10.20 Rice Paddy Prone.

Figure 10.21A Elbows inside knees.

Figure 10.21B Rear view of elbows being pushed tight by shooters knees.

Sitting Positions

Getting into a good sitting position will require twice as much time to assume as other positions. However, the sitting position is much more durable, which, in turn, means, you will be able to stay in the position for quite some time without serious fatigue. If the situation permits and your gear does not hinder the assumption of this position, then do it.

Conventional Sitting Position

The first sitting position is a Conventional Sitting Position. This position is built by sitting cross legged or Indian style, with your butt on the ground. Once in this position, you will place your elbows inside the bends of your knees. See Figure 10.22. It is very important to ensure that you do not have bone on bone contact. Always try to place muscle on bone. For example, place your elbow (bone) on the soft portion inside of the bent knee. You may also lean forward slightly farther and place your triceps on the front side of your knees. See Figure 10.23A-B. This will be especially difficult if you are wearing a lot of assault kit. It may also be difficult if you have too much belly. While in the conventional sitting position, experiment with other cross legged positions such as crossing your lower legs or ankles. If you are limber enough, this is an excellent position. One other position while seated is to slide your biceps across the front of your legs, but slide them so far forward that your elbow actually sits on the side of your foot. See Figure 10.24. Almost as if you were shooting prone, breathing can be an issue in this position so be ready to break your position slightly if you need to suck up some more sight cleansing oxygen.

Figure 10.22 Elbows on inside of knees (Conventional Sitting)

Figure 10.23A Elbows in front of knees (Conventional Sitting)

Figure 10.23B Front view of elbows in front of knees.

Figure 10.24 Elbow on side of foot (Conventional Sitting)

Spread Legged Sitting

If having too much combat gear, or possibly a front mounted rucksack (belly), gets in the way of a conventional sitting position, then you may have to adapt your position to overcome these hurdles. Simply spread your legs with your knees bent and build the best possible sitting position you can. See Figure 10.44A. You may also be able to keep your ankles crossed when you spread out into this position. Experiment and see what works best for you. See Figure 10.44B.

Figure 10.44A Spread legged sitting position.

Figure 10.44B Crossed ankle sitting position.

Front Knee Up Sitting Position or Rocking Chair

The next sitting position is the Front Knee Up or Rocking Chair, Sitting Position. This position will work well if you want to be comfortable but do not want your feet falling asleep, as they sometimes do, in the conventional sitting position. This position also works well when shooting over the parapet of a rooftop. This position can be used whenever you want the stability of the sitting position but need more muzzle elevation. This position is also more durable than the kneeling position.

To get into the position, drop to your butt, as with the previous sitting position. Once seated, lift your front knee, or left knee if a right handed shooter. Once the knee is up, place the sole of your left foot flat on the ground. Try to keep it completely in contact with the ground. If you are up on the heel you will have difficulty keeping a steady position. With the left knee up and the foot flat on the ground, lay your right ankle across the top of your left foot. Once the right leg has been positioned, let it sag under its own weight, you want to eliminate as much muscle tension as possible. Your right knee should be free hanging. Now place the weapon across the inside of your left knee, if you do not have a sling this position will be very difficult. See Figure 10.26A-B. If you do not have a sling that allows quick adjustment you will also have a difficult time. Once the hand guard is positioned on the right side of the leg, grasp the magazine and your knee with the left hand. This will be different depending how you are built. The bottom line is that when you are in position, your sling should be tight across your back, which will allow you to slightly lean back into the sling without tipping over. This position works very well if you are sitting against a tree or side of a building. And, if necessary, you can sit back to back with your mate. I know this may sound crazy, but sometimes you have to do whatever it takes. See Figure 10.27. When you fire the weapon, do not fight the recoil. Let it rock you back slightly. If you are relaxed and the right leg is crossed across your left foot, and it is relaxed as well, your body will tilt and recover on its own. If you attempt this position with a high recoiling weapon, ensure you pay close attention to your scope to eye distance; it will bite you if you are not careful.

Figure 10.26A Weapon inside of knee. Rocking Chair Position.

Figure 10.26B Rocking Chair, viewed from the other side. Sling must be tight to improve this position.

Figure 10.45A Grabbing Grass to stabilize position.

Figure 10.27 Back to tree

Figure 10.45B Close up of grass pulled into position.

Stacking Your Feet

The next sitting position looks absolutely crazy, but it works very well for some. Once I figured out this position, I was hooked. It is called Stacking Your Feet, or at least that is what I call it. This position has very limited application, but I have used it to shoot deer from a rather steep hillside. Once again, to get into the position, drop to your butt. Instead of crossing your legs, you will extend your right leg in front of you, toward the target. The heel of your left foot will sit on top of the right toe, which will be slightly bent back towards you to create a shelf for the heel. See Figure 10.28. You may need to bend your knees slightly to bring the feet in close enough to rest

> **Grab some GRASS.** Once upon a time in a land far away, I was told to grab some grass while in a shooting position to help stabilize my sights while shooting. As you can imagine, I either thought this fellow had read too many bed time stories, or he had smoked some of that which he wanted me to grab. I tried it anyway, and believe it or not it works. So the next time you are in a position. Reach over and grab a handful of grass and pull it into your position, you will be amazed at the help it gives. See Figure 10.45A-B.

your rifle on them. This is one of those positions where either you have it or you don't. Give it a try. Be careful getting your rifle in position. You should not have your fingers anywhere near the muzzle of the weapon for obvious reasons. I sometimes lay my left hand, palm down, almost flat on the bottom of my left foot for grip, then I grip the rifle with a couple of fingers, or I grip my forward pistol grip. If possible, you may try hooking your pistol grip on the bottom of your foot. See Figure 10.29 Hooking pistol grip to bottom of foot. What a weird position, but it works. When you are in this position your field of fire is very limited, so establish a good Natural Point of Aim, practice building the position to limit movement once the position is established. While in the position, if you can apply slight pressure to either side of your magazine or the rifle, it will make you slightly more stable. This slight pressure coming from the inside of a forearm or a light wrist touch can make a noticeable difference on your sights. Any bit of help in quieting the sights is a plus.

If you realize that you need to change the elevation of the front of the weapon, you can simply rotate your feet to change the elevation. This can be done while in position, just watch your sights as you rotate your feet. You will end up with one foot almost flat. Keep attempting the position, you may be able to come up with your own spin on the situation.

Figure 10.28 Stacked Feet

Figure 10.29 Hooking pistol grip to bottom of foot

Figure 10.30A Shooting without pistol grip, using thumb hooked into sling.

Figure 10.30B Shooting with thumb hooked through sling.

Figure 10.30C Shooting without pistol grip, clamping weapon to foot with support hand.

Figure 10.31 Stacked feet with bottom foot almost flat

Kneeling Positions

If sitting or prone are not feasible, you should always try to stay as low as possible. Kneeling is quick and provides a somewhat more stable platform, definitely more stable than standing. Kneeling is going to be the most popular or widely used tool in the box. You can use kneeling for corner clears, while conducting a tactical reload or while shooting around the side of a car. If you need a quick, but steady position while moving towards a target, you can quickly drop into the kneeling position.

Conventional Kneeling

The first kneeling position we will talk about is Conventional Kneeling. Conventional kneeling works well during target engagements. But, normally for tactical shooting, it is just a starting point to build a somewhat better kneeling position. For Conventional kneeling, place your rear or your strong side knee on the ground. Leave the front knee or support side knee in the upright and bent position. See Figure 10.32. Always try to make contact with the knee and elbow of the support side in some form or fashion. If you must make bone on bone contact to get the shot off then that is acceptable, however you will be more stable and the position will not shift as you shoot if you are able to get the meaty portion of the tricep to rest on the knee instead. See Figure 10.33.

Figure 10.32 Conventional Kneeling Position.

Figure 10.33 Elbow in front of knee.

Figure 10.34A Seated Kneeling Position.

Seated Kneeling Position

A modified Kneeling position is the Seated Kneeling Position. This position is very similar to Conventional Kneeling, with one small exception: You will rotate your strong side foot inboard and sit down on your flattened foot. This position does not work well if you have bad knees or are not very flexible. See Figure 10.34A-B Seated Kneeling. If you can sit down, it will be a much more stable position. It will require a little bit longer to recover to the standing position though, so this should be considered. If you are able to get into this position, you may also be able to rest your rifle on the front knee or wrap your arm around the knee to make the position slightly more stable. If you wrap around the knee, grasp your shirt or jacket with the non firing hand and compress the whole position. See Figure 10.35A-D. If you are shooting a heavy recoil weapon, be careful not to get to close to your optic. This will be a tight position and getting hit with the scope is quite possible.

Figure 10.34B Seated Kneeling, second view.

Figure 10.35A Magazine and pistol grip stacked on shooters knee.

Figure 10.35B Wrapping arm around front of magazine.

Figure 10.35C Grasping shirt with support hand to tighten position.

Figure 10.35D Be prepared for the weapon to tap you on the glasses if you don't have a tight position.

Stretch Kneeling

Stretch Kneeling is another modified version of the Conventional Kneeling position. This position will allow you to quickly establish a somewhat stable position for mid range target engagements. To build the Stretch Kneeling position, take a step forward as you go into position. Stretch your rear leg or strong side leg as you get into the position. Your tricep on the support arm should drop down to make contact with the knee. See Figure 10.36. This position stretches your groin muscle as well as the quadriceps. Because of this stretching, you will not want to be in this position for a long time. Before you practice this technique, stretch a bit, just to make sure nothing gets overstretched. If you want to get every little bit from this position, slightly rotate your support side elbow under the weapon and touch the magazine to the inside of your forearm. It will help. See Figure 10.37. Stretch Kneeling is very quick, and much more stable than offhand. Simply another tool for the box.

Figure 10.36 Stretch Kneeling.

Figure 10.37 Elbow rotated to touch magazine.

Kneeling Around Corners

You will need to modify the kneeling position when preparing to kneel around an object, whether it is a brick wall or a tree. For Kneeling Around Corners, one important change to your technique will be which knee is up. Your rear knee is generally on the ground for a kneeling position; however, when kneeling around corners, you should leave the rear knee up. Since you have an obstacle in front of your position, use it for support of the front of the weapon. Drop the front leg, raise the rear leg and use it for support of the strong elbow. You will also need to make sure the support hand is able to keep the weapon from sliding down the wall as you fire. This can be accomplished by using your hand in the shape of an "L." Or you can actually grab the tube in your non firing hand. See Figure 10.38A-F. This also works when conducting a dynamic corner clear. You are able to step more aggressively into position with the rear leg in the upright position. This technique only applies if you are shooting right hand around a left corner or left hand around a right corner. It will take a while to get used to getting into this position. Ever since you were a little lad playing with green Army men, the rear knee has been down. So take your time, think about what you are trying to accomplish, and practice, practice, practice. This position will change the way you are able to engage from around a corner drastically. It is also of importance to point out here that you need to be able to engage from the right or left side of the barricade. See Figure 10.38G-H.

Figure 10.38A Kneeling around a corner.

Figure 10.38B Rear view.

Figure 10.38C Applying pressure with support hand. Support hand in an L shape.

Figure 10.38D As you get lower, continue to apply pressure to back of arm with knee.

Figure 10.38E Low kneeling, holding rifle tightly to barricade.

Figure 10.38F Rear view of low kneeling.

Left Handed Kneeling

Figure 10.38G Left handed kneeling around a corner.

Figure 10.38H Left handed low kneeling.

If you come upon a somewhat stable object that can provide support, use it. If you have a small tree, mailbox, or light pole, use it. Take up the Kneeling Around Corner position, with one small exception. Wrap your support arm around the pole and hug the post as you build the position. See Figure 10.39. Maybe beside a fire hydrant. This will also allow you to maintain the position quite a bit longer. If you have an adjustable sling, this will be a good time to adjust it for the position. If the object you are using will provide some bullet stopping capability, then that is a plus. If not, take the shot quickly to eliminate the threat and move on.

Figure 10.39A Kneeling with tree or pole.

Figure 10.39B Support hand pushing rifle against pole.

Figure 10.39C Opposite view. Note how sling is tightened as the position is tightened.

Standing Position around a corner

One position often overlooked when discussing positions is the standing position. There are many times that there are confined space situations where the shooter doesn't have enough room to attain a more stable shooting platform or you just may want to quickly clear a corner or pull security from a standing position. This position is easy to build since we are used to standing and shooting; but, there are ways to make this position work better in a tactical environment. First, you should use your cover or concealment if possible to help stabilize the position. Secondly, leg positioning is crucial for a solid and safe position. I say safe, because some positions will allow the shooter to fall or step into the line of fire quite easily, whereas a slightly modified position will keep you behind your cover even if you are bumped. I prefer to square my body and stance a little bit more than normal for clearing these corners. With a square stance you can keep your body more stable, and if you are bumped you can simply lift your leg to regain balance and pull yourself back into position. See Figure 10.40A-B. If you are shooting around the corner with a bladed position, then the only way to get back behind cover is to step into the danger zone, or line of fire. See Figure 10.40C-D.

Figure 10.40A Squared position for shooting around a corner.

Figure 10.40C Bladed body position around a corner.

Figure 10.40B If bumped, simply lift your leg to move back behind cover.

Figure 10.40D If your balance is lost or you are bumped, you will have to step into the danger area before being able to step back.

Junkyard Prone

Shooting over cars. Why would you shoot over a car when you can shoot under? Not sure that requires an answer. If you can shoot under, then by all means do it. If the situation dictates that you must shoot over, you need to know how to successfully accomplish this. One of the downfalls of shooting over the hood of a car will either be the lack of stability or the lack of cover when you attain a stable position. To remedy this, I have started using a version of the SBU prone. I call it Junkyard Prone. Junkyard Prone allows you to maintain very good cover while engaging targets over the hood of a vehicle. See Figure 10.19A-C. You will lose your stock to cheek weld, and will more than likely not have a great amount of shoulder on the stock either. Don't be alarmed. Just put your sight on target and squeeze the trigger. The only caution for this position is creasing or putting a hole in the hood of the vehicle. When practicing, make sure you don't mind powder burns on the vehicle you are shooting over. Also, keep the muzzle elevated enough to miss the apex of the hood. Normally if you can see the target from this position, so can the bullet. This is due to the weapon laying on the side and not needing large amounts of adjustment to clear the hood. If you do hit the hood in a gunfight, you will more than likely not notice until it is over. I have shot several cars to see what happens and I have never received frag from hitting this portion of the vehicle. Another note of caution is that I have cracked windshields shooting this position while using a compensator. Flash hiders have never presented an issue, but the extra jetted gases from a muzzle break tend to cause a lot of damage. Compensators will even cause the paint to peel if you aren't careful.

Figure 10.19A Normal shooting position behind vehicle with Vertical Fore Grip.

Figure 10.19B Shooting over vehicle without vertical grip. Notice amount of shooters head that is still exposed.

Figure 10.19C Junkyard Prone.

Figure 10.19D Side view of Junkyard Prone.

Rollover Kneeling

At some point in time you may need to shoot a kneeling version of rollover prone. See Figure 10.42. The only difference is that you will be in a kneeling position, possibly to shoot through a fence or some sort of horizontal barred gate. If you are able to rest your hand guard on the obstacle, then go ahead and raise the rear leg, or strong side leg. Next roll your upper body over until your sights and muzzle are clear of the obstacle. If you roll over and the rear leg is up, it will provide excellent support for your elbow or upper body. Just remember what happens to your trajectory when you roll your weapon to either side.

Figure 10.42 Rollover Kneeling. Using rear knee for support.

Broke Back Prone

Broke Back Prone is a modified version of Rollover Prone. Broke Back Prone allows the tactical shooter to assume a rollover prone position when wearing a lot of gear, or if you need to have the mobility to quickly move in and out of your position. I was trying to come up with a name for the position. All it took to name this position was to show a few Law Officers in a class in San Jose, CA. They named it Broke Back Prone. I thought the name fit, so there you have it. As you can see in Figure 10.10A-B Broke Back Prone, the name fits. This position will also allow for quickly clearing under vehicles without getting completely prone. It may be a funny position with a odd name, but it works.

Figure 10.10A Broke Back Mountain Prone.

11. MEAT AND POTATOES
Target to target acquisition

After training on the range to make the perfect shot, reading your sights, squeezing the trigger and following through to get a hit on target, then it is time for multiple hits on several targets. This is where the rubber meets the road. One shot, one kill fits nicely for the Sniper community; however, for a Tactical shooter, that first shot may just get things rolling. Regardless of your choice of weapon system-pistol, rifle, or shotgun- acquiring and engaging multiple targets quickly can be the difference between success and failure.

Lead with your eyes

Quick target acquisition is very visual. Before you can shoot a target, you must look at the target. From a young age we have always looked, then pointed. For example, "hey, check out that wicked car" you look and then point. You never point and then look. The same can be said for shooting multiple targets. You should be looking for the next threat or target, once you see it, quickly snap your weapon in that direction. Don't get tunnel vision just because you are using a scope! It is even more important with a scoped rifle due to your decreased field of view. Looking over or around your scope will let you identify visually. Next, start the weapon moving. Keep your eye on the target and let the weapon's sights come into view. There is no need to move the weapon slowly between targets. As I said earlier, "take your time shooting, don't waste time on anything else." As the sight arrives in the general vicinity of the target, it is time to practice your visual patience and make this shot count. Once you have acquired your sights, squeezed the trigger, and followed through calling a good shot, simply repeat the process.

Once again, look for a target with your eyes, not through your sights. Identify the target. Is the individual a threat? If so, drive the weapon to that target quickly, acquire sights, trigger squeeze, follow through, call the shot and repeat as necessary. See Figure 11.1A-F. Leading with your eyes. You will find with practice, this technique will become second nature. Leading with your eyes and not your weapon also allows you a split second lead before your sights get there to discern if this is in fact a shoot target, or instead just an innocent bystander caught in the wrong place at the wrong time.

Lead with your eyes

Figure 11.1A Start with eyes and weapon on the same target.

Figure 11.1B Aggressive shooting stance, preparing to snap the weapon to the new target.

Figure 11.1C Eyes lead to the new target. Don't get tunnel vision.

Figure 11.1D Eyes are on the new target. You have followed through on your last shot.

Figure 11.1E Eyes remain on the target as you drive the weapon to catch up with your eyes.

Figure 11.1F If your eyes are on the target, your weapon will stop on your eyes.

This eye leading technique becomes more and more important when you are in a close quarters environment. You must not get glued to your sights. This will help you open up your tunnel vision, helping you to see target indicators you would normally miss when glued to your sights. Your eyes are amazing tools; make good use of them. If your eyes lead, your gun will follow.

Is Smooth Fast?

Often times when I am on the range I will hear shooters say "smooth is fast." This statement is very true however they leave out the important ending. "Smooth is fast, but slow is just fricken slow." In the end, the only thing that really matters is placing the bullet on target. Whether it is a piece of paper or human flesh, making a precise hit is what counts.

People tend to lollygag on the draw of the pistol, the rifle mount (presentation), or weapon movement between targets. However, when they start shooting, the race is on. Why? Most of the time, this is caused by shooters not having visual patience. We don't like waiting for the sights to become properly aligned on the target, squeeze the trigger and lastly, follow through. You need to be in a hurry to acquire the target and quickly place your sights on the target. Slow down and make the shot count!!

Shaving time on your draw or long gun presentation is great; but, in the end, rushing the shot negates any time you may have saved up to this point.

If you do break the shot early, as long as you have squeezed the trigger and followed though, then you will know immediately that it was a bad shot. Now you can accurately call your miss, reacquire the target, and shoot again.

You will also need to learn to accept some sight float. Allowing the sights to float around the vitals while continually making small corrections as you squeeze the trigger is paramount. If you simply wait for the perfect sight alignment with the target, you will be extremely slow, and in this business slow usually results in bad things happening to the good guys. If you start the trigger squeeze while continuing with sight refinement, you will have a large percentage of your shots fall into the vital zone. Your eyes will continue to adjust and if the sights become too unacceptable your finger will stop squeezing. This specifically applies to your target disappearing or becoming obscured by a no shoot/hostage target. You must always be aware of the changing tactical situation.

A simple drill to practice your eye leading techniques is as follows: Set up two targets that can be easily seen in your periphery. This translates into being able to see one target when your weapon and eyes are on the other target. You must practice this drill with both eyes open. You should start out with the targets at the 10 yard line about 5 yards apart. See Figure 11.2. Place your sights and eyes on one target. Snap your eyes to the second target when you are ready to begin. Once your eyes are there, drive your weapon quickly to catch up with your eyes. After the sights settle on the target, start your trigger squeeze, insuring that you have sufficient sight alignment. When the shot breaks, follow through and call the shot. Is the shot where you said it would be? Once this target is neutralized, snap your eyes back to the first target. When your eyes arrive, drive the weapon again to catch your eyes. Complete the previous cycle of events. You must make sure you are calling your shots.

Are you over swinging the target with your sights? If so, practice placing your eyes on the target, snap the weapon to your eyes, and dry fire to see how it feels. This technique takes a lot of practice, but once you become comfortable, you will be able to not only shoot faster, but to target discriminate much quicker than if you were looking through your sights all the time.

Driving your weapon drill

Figure 11.2C Drive the weapon back to the right target to catch up with your eyes. Engage the target. Repeat as necessary.

Shooting is extremely visual and you must always update information as you receive it. You must be ready to react. The quicker you react, the farther ahead of the enemy you will be. And this is a race. A race you don't want to lose.

"Make the Shot Count"

Figure 11.2A Start by engaging the right target. Move your eyes to the left target.

Figure 11.2B Drive the weapon to the left target. Engage the target. Once you have called a good shot, move your eyes back to the right target.

12. IS WEAK REALLY WEAK?

Shooting from your support side

In the tactical community, we often become lulled into a false sense of security when it comes to weapon manipulation and marksmanship. This false sense of comfort is very evident when tactical shooters are asked to do something which they think is off the wall. For example: shooting a rifle or shotgun from your support side. In almost every tactical situation I can think of, it is important to have the ability to successfully engage targets from your support or weak side will give you a huge advantage. This, in turn, increases your chances of survival. We as shooters have even labeled this shooting as weak!! Weak side, weak hand. To be ready for the day a real gunfight occurs, you need to be mentally and physically prepared. If you are physically able to shoot from your support side and practice this ability persistently, then your ability will improve. With physical success will come mental confidence. If you are mentally and physically confident in your ability, then you will not have to take the extra time in a gunfight to push doubt out of your brain housing group. Removing doubt will help speed your reaction to act in a life or death confrontations.

Taking a little time to add support side skills to your skill set is a logical decision as far as I am concerned. Shooters who have attended any of our rifle courses will be the first to tell you that these skills are easily attained with very little practice. Once you have the basic skills and understand when, and when not to apply these skills, you will become a better, well-rounded tactical shooter.

There were those who said red dot sights are bad. They were wrong.

There are those who said using a 5.56 caliber rifle for CQB (close quarter battle) will have over-penetration issues. They were wrong.

And, of course, there are those who say you should never transition to the support side shoulder for tactical applications. And you guessed it, they are wrong as well.

If you plan to use the carbine to its utmost potential, then you must be able to employ the weapon system in any situation. If you are riding in the right front seat of a vehicle and those around you expect you to cover a sector of fire, you must be able to switch to the left handed shooting position to fulfill these duties. Enough said about that.

Eye Dominance

Normally when I ask shooters to try these techniques, I get a lot of excuses. One common excuse concerns eye dominance. Nine times out of ten this is just a shooter being lazy. I have a close friend who lost an eye in an Army training accident. I have never, ever, heard him say "I can't." He had to put his optic on a slightly elevated rail so he is able to shoot right handed with his left eye. The only acceptable excuse is that you are blind in one eye and can't see out of the other and if this is the case, you're out of luck.

When you finally decide to become a true tactical shooter and practice with your rifle and pistol using your support side, here are a couple of precautions. Always practice new drills with an unloaded weapon. This will allow you to gain confidence with the manipulation of the weapon with your support side before working on the marksmanship skills with the support hand. Also, ALWAYS engage your safety between strong to support side transitions. ALWAYS. A mishandled, unsafe weapon could have dire consequences. Be Safe!

Don't just practice shooting. Practice clearing malfunctions with one hand, strong and support side. Lastly try these drills at night during limited visibility.

Support side shooting skills are not just for close quarter engagements. Once trained, you should not have any issues successfully engaging targets with your rifle at every distance and in all positions, strong or support side.

> *While fighting in Somalia in 1993, soldiers who had severe wounds to their strong side arms had to switch to their support side and engage targets from varying distances. With only the use of their support side, these soldiers learned on the battle field. It is my goal to teach shooters to have these skills ready before the day that it is necessary to utilize them occurs.*

When you train, train hard. Practice the most difficult scenarios, practice doesn't make perfect, but it will help ingrain reflex actions for stressful situations! And just in case you didn't know this, gunfights are stressful!

Transferring your carbine from strong side to support side in a tactical situation

Whether you are moving through the jungle, scampering from light post to light post in the urban sprawl, or moving down a hallway in an abandoned building, if you are able to quickly and confidently transition your carbine from strong side or dominant side to your support shoulder, then you will have a tactical advantage.

Once you have taken the time to gain confidence with your rifle fired from your support side during marksmanship, flat range training, then it is time to develop skills to fluidly move the weapon between these two positions.

First of all, you should have a sling that will easily allow you to conduct these transitions while retaining your weapon. The following procedures will be executed with the Viking Tactics quick adjust sling. No matter which sling you use, figure out the best means to your end.

Secondly, ensure you always start these drills with your weapon on safe. Your weapon's safety should always be on, unless you are engaging a threat target. Develop a technique that works for you to disengage the safety once you are ready to acquire a target from your support shoulder. This technique should also allow for sound weapons retention.

Safety Manipulation

There are two techniques that I use for disengaging and re engaging the safety on the AR weapon when using the left hand.

First let me discuss how not to manipulate the safety. I tend to see a lot of left handed shooters use their firing hand thumb to accomplish this task. It does work pretty well, but it does not pass the no- nonsense test. First you lose weapon retention capabilities, or they are sharply diminished. This is due to the fact that you can't fully hang on to the rifle's pistol grip while conducting this manipulation. Secondly, it takes time to get your thumb back around to the other side of the pistol grip before shooting. See Figure 9.14.

There are two techniques I use for left handed safety manipulation.

The first technique is the use of the trigger finger for this task. This seems to work well for most, but it does have a shortcoming. When using this method, it takes a bit longer to get your finger back on the trigger. Re-engaging the safety with your index finger works well, especially since you are not in a hurry to engage targets at this juncture.

The next technique is the use of the knuckle of the left hand. Not the finger joint, but the knuckle. This will give you plenty of power to force the safety into the off position. It also allows for good weapons retention capabilities. See Figure 9.15A-C showing the correct technique for disengaging your safety.

Figure 9.14A Disengaging the safety with the support side thumb.

Figure 9.14B Not a good grip for weapons retention.

Figure 9.15A Index finger of support hand in position to deactivate the safety.

Figure 9.15B Sweeping the safety off with the support index finger as you mount the weapon.

Figure 9.15C Once the safety is off, you are ready to fire.

Transferring from Strong to Support Shoulder

Now, with your weapon on safe and your sling in its normal carry configuration, we will talk through the steps one at a time. If you are left handed, or normally fire from your left shoulder, these directions will be completely backwards. Your sling, when carried in the low ready, will go from the butt of the weapon, over your right shoulder, across your back and under your left arm, finally connected to the carbine somewhere along the left side at least midway of your rifles fore stock. See Figure 5B.7 for proper wearing of the VTAC sling.

Figure 5B.7 VTAC Sling

Once configured in this manner, to start a smooth transition you will need enough slack in your sling to be able to slide your left arm between your rifle and sling. See Figure 12.1A-D. With this completed the sling will now only be around the back of your neck. Next, grasp the fore stock of your rifle with your left hand. See Figure 12.2. Grasp far enough forward so you will be able to control the full weight of the weapon with your left hand. You are now ready to start the actual transition.

Figure 12.1A Start with weapon in normal shooting position.

Figure 12.1B Grasp the quick release on the VTAC sling.

Figure 12.1C Loosen the VTAC sling by simply pulling the release lanyard.

Figure 12.1D Drop support side elbow through sling.

Figure 12.2 Re-grasping the fore grip of the weapon.

With your firing hand, reach forward and grasp the front of the magazine well securely. See Figure 12.3. Now in one fluid motion, pull the weapon away from your right shoulder with your right hand, while transitioning your left hand to the rear pistol grip. You will also need to take a slight step forward with your strong side leg to allow the rifle to be placed in a position that allows for a strong fighting stance. See Figure 12.4-12.5C. As you switch hands you should also be moving the weapon to your left shoulder. Once the rifle is securely in the pocket of your left shoulder, you should be ready to engage a target if the opportunity arises. Your strong and support side shooting positions should mirror each other. I know this seems simple but I see shooters with whacky support side positions all the time. Once they know what they are doing, it is easily remedied. When you make the switch your sling will become crossed on your neck. This should not hinder your ability to attain a good position. If it does, loosen your sling and try again. Better to find out now, than in a life or death experience. It is a 50/50 situation, so don't end up taking a dirt nap.

Figure 12.5A Grasp the pistol grip with the support hand.

Figure 12.3 Grasp the front of the magazine with the shooting hand.

Figure 12.5B Place Strong side hand in a forward position to allow use of your light, and to get hand clear of the ejection port.

Figure 12.4 Transition the weapon to the support shoulder. Notice sling will be crossed at neck.

Figure 12.5C Finish mounting the weapon, deactivate safety, and engage threats as necessary.

Transitioning from Support Side to Strong Side

When the situation changes and it is time to go back to your dominant shoulder, simply repeat the process with opposite hands. First things first, make absolutely sure the weapon has been placed back in the safe position. Figure 12.6, left hand, which is now your firing hand, grasps the magazine well, See Figure 12.7, right hand grabs the pistol grip as you place the rifle back into your right shoulder. See Figure 12.8-12.9B. This process will become second nature with a little practice. A good drill is to start with your weapon slung, engage two targets right hand, two left, two right and two left. This will help to build confidence, *always place the weapon on safe between shoulders.*

Figure 12.7 With your left hand grasping the magazine well.

Figure 12.8 Place rifle back into strong side shoulder.

Figure 12.6 Ensure that the weapon is on safe.

Figure 12.9A Grasp the pistol grip with the strong side hand.

Figure 12.9B Move support hand forward to be in a position to activate weapon light.

Keep your eye on the target

What is your head doing or where are you looking during this transition? Keep your eyes on the area where a threat may present itself, if possible. At night you will not be able to see your weapon as you move into position to shoot from your support side. Granted if you had a target to engage you would shoot, not transfer the weapon. That is unless you really want to make this sporting!

Eye Dominance Tricks

If while practicing your support side shooting skills you continue to have eye dominance issues, or are simply having a hard time closing the normally dominant eye, here is a little trick. As you move into the support side shooting position, you will normally be behind cover or concealment, if not why would we shoot support side? As you take up the position, try to set your body close enough to allow the target to be obscured by the cover item for your non firing eye. In other words, get as close as possible to the wall, your eye will not be able to see the target. This will allow your other eye, the normally non dominant eye, to take over and finally do its job. There are other tricks to help train the non dominant eye for support side skills. One such technique is placing a small piece of transparent tape across the top portion of the non firing eye glass. Just enough to obscure the target yet allow light into the eye. This is for training purposes only. Don't rely on this trick; it is not for use when the time comes to fire your weapon in anger. Use it sparingly for training and as soon as you feel you can make the shot without the tape, get rid of it.

Practicing Your Support Side Transitions

When you have mastered the rifle transfer on the flat range, try using this technique with barricades, around corners, and near other normal obstacles, such as cars. If you set yourself back a couple of steps from a barricade and start with the weapon in your normal firing grip, it is great practice to move into the kneeling corner clear using the support side. You may feel disjointed at first, but with minimal practice it will become second nature. You must make sure you are still using the correct knee for support. If the kneeling position is from a barricade, then the front knee will be down and the back knee will be up, providing support for the support side elbow. See Figure 12.10A-M Shooting barricade strong side, support side and back to the strong side.

Figure 12.10A Start a step away from the barricade.

Figure 12.10E Grab the front of the magazine well with your strong hand, then drop your support side elbow through your sling.

Figure 12.10B Assume a good kneeling position.

Figure 12.10F Hand the weapon to your support side.

Figure 12.10C Engage a target from your strong side.

Figure 12.10G Support hand is in a position to receive the weapons pistol grip.

Figure 12.10D Place weapon on safe and start your transition.

Figure 12.10H Slide strong side hand forward to help with forward grip of the weapon and engage your target. Note elbow resting on rear knee.

Figure 12.10I Place weapon on safe and grab front of the magazine well with the firing hand, which is now your left hand.

Figure 12.10J Hand the weapon back to the strong side.

Figure 12.10K Slide back to the other side of the barricade.

Figure 12.10L Build a good kneeling shooting position.

Figure 12.10M Engage the target again from the right side of the barricade.

Tools For the Toolbox

Support side shooting is one of the most valuable tools you can add to your Tactical Toolbox. Even though it is highly important, little work has been accomplished in the community regarding support side shooting. The reasons it is often over-looked are simple; either you have instructors who are not competent shooting support side or you have instructors who have never been shot at. If you have been shot at, you will try to get as small as possible. These support side shooting positions allow you to not only present less of a target, but, return effective fire. As a Tactical Shooter, you do not want to have a WEAK side!

13. FACING MOVEMENTS

Often times when asked to conduct simple facing movements, shooters end up looking like they have equilibrium issues. They stumble and stammer; they completely forget how their feet work. When we add shooting to the problem, watch out.

To eliminate this issue during a high stress situation that we must practice over and over, ad nauseam. You also need to develop moving and turning techniques that are comparable to your normal movement activities. Keep it simple to reduce the training time needed to become comfortable and more importantly, to be safe.

When you are practicing turns, start with the absolute easiest turn, turning from the ready position to your support side. In other words, face to the right side of your range if you are right handed, and turning to the left to engage the target. Always dry fire these techniques several times before sending lead down range. So, with a target to your left, assume a good shooting stance as if you were planning to shoot straight forward, which would be the right side of the range. See Figure 13.1 Your carbine should be in a low ready position, finger off the trigger, weapon on safe, and your weight on the balls of your feet. Upon the decision to turn, you will lead with your eyes. This is the most important movement since if your eyes are not on the target, then you don't have a chance of hitting the threat. Leading with your eyes will also help you begin the target discrimination process (determining if the target is in fact a threat) before your sights are even on the target. This will also require a head movement. Once you have rotated your head enough to acquire the target with your eyes, See Figure 13.2. The rest of your body will start to move. Do not raise your rifle as you turn for three reasons; first, it will cause an unsafe situation if you are in close proximity of your team mates; secondly, you cannot turn as quickly with your rifle in the fire position; and lastly, you will not be able to stop the rifles momentum as easily if you have the weight of the rifle extended to your front. Your weight will shift to the ball of your left foot as you push off with your right foot. See Figure 13.3A-B. This will allow you to make one rapid, smooth movement to come face to face with the target to be engaged. Once your body is in line with the target, quickly drive the rifle to the center of the target. As I said before, bring the weapon to you and don't drop your head to the weapon. See Figure 13.4A. To reiterate, once you have a target indicator (ie sound, movement, hair on the back of your neck stands up), your first step is to turn your head and eyes toward the target. Next step, forward with your right foot, pivoting on the left foot, driving the rifle to the center of the target once your turning movement is complete. See Figure 13.4B Once you have engaged the target, follow through and assess the situation.

Right handed shooter turning left

Figure 13.1 Start position for turning to the left.

Figure 13.2 Head turning towards the target.

Figure 13.3A Right foot steps forward as your body turns toward the target.

Figure 13.3B Body ends up square to the target. Not muzzle is still in a down position.

Figure 13.4A Mount the weapon as you finish your turn.

Figure 13.4B Engage the target.

Left handed shooter turning right

Figure 13.5 Start position for left handed shooter turning right.

Figure 13.6 Head turning towards target for left handed shooter.

Figure 13.7A Make an aggressive step forward with the left foot.

Figure 13.7B Keep the muzzle of your weapon down until you have completed the turn.

Figure 13.8A Once the turn is complete, drive the weapon aggressively to the center of the target.

Figure 13.8B Engage the target once your sights have settled on the threat.

Never step backwards when conducting facing movements in a tactical environment. To better understand this, picture yourself in a hallway of a building hit by artillery fire, bricks and debris on the floor. As you move down the hall you hear something go bump in the night. You turn and decide to step to the rear as you make the turn. Murphy, being the man he is, will be there to put a brick behind your foot. Normally, if you lose your balance to the rear, there is no recovering. Fighting from your kiester is not a sound tactical decision. Therefore, always step forward. In this situation if you had chosen to step forward towards the threat, you may stumble, but normally you will be able to catch yourself. The worst case is to be a short drop to a kneeling position, a position that will still allow you to send rounds toward the threat.

Turning to the rear 180 Degrees

180 degree turns require a little more thought. First you will have to decide which way is the best way to turn. The rule of thumb I use is to turn in the direction my head initially turns. If this is to the right, my body will follow. I will walk you through turns to the rear in both directions. Remember the forward movement rule and you will be fine; this rule will also help you to remember the correct technique.

*Note all descriptions are for a right handed shooter.

When conducting range fire, always incorporate turns if range rules and set up will allow.

180 Degree Turn to the Left

Start facing up range, in a proper shooting stance with your weapon safely pointed towards the ground. See Figure 13.9. Once you are ready, turn your head to the left as far as you can. If you can't see your target yet, don't worry about it. See Figure 13.10. As you start the turn you will be able to get your head in the position that you need to. Your weight should be on the ball of your left foot. Push off with your right. Try to push hard enough to rotate your body fully with one step. If you don't, you may have to stutter step to get into the preferred shooting position. See Figure 13.11-13.12. Once your body has rotated, drive the weapon up into the center of the kill zone, remove the safety, and squeeze the trigger. See Figure 13.13. Follow through and try it again, ensure that the weapon is placed on safe as you come off target. If you feel like you are stepping away from the target, then you are correct. The key is a forward move. It will feel awkward at first, but if you are stepping forward, you are on the right path. When turning left you will end up slightly farther away from the target than when you started, but you are still making your movement in a forward direction. If this sounds confusing, try it a few times and you will understand. You should be able to get your eyes on your target ahead of your rifle in plenty of time to target discriminate.

Right Handed shooter conducting a turn to the left rear

Figure 13.9 Start position- facing to the rear, turning to the left.

Figure 13.10 Head starts the turn, always leading with our eyes.

Figure 13.11 Right foot steps aggressively forward, ensuring that we are able to turn far enough to engage the threat.

Figure 13.12 Finish your turn before mounting the weapon.

Figure 13.16 Right foot stepping aggressively forward. Rotating on the ball of the left foot.

Figure 13.13 Drive the weapon to the center of the target.

Figure 13.17 Finishing the turn, weapon still in the low ready position.

Left handed shooter conducting a turn to the left rear

Figure 13.14 Start position for a left handed shooter-facing to the rear turning to the left.

Figure 13.18 Drive the weapon to the center of the target and engage.

Figure 13.15 Head starting to turn to the left.

Facing to the Rear- Turning to the Right

Once again begin facing up range, weapon on safe, finger off the trigger. On the start signal or indicator, turn your head to the right, trying to acquire the target, at least in your periphery. See Figure 13.19. If you are unable to see the target, don't worry.

You will see it as soon as your body begins the turn. You will transfer your weight from the left foot to the right as you begin your aggressive turn. Normally when shooting a rifle, the majority of weight is on your front foot; therefore, you will have to conduct the weight transfer I am addressing now. Once the turn begins, your weight will go to the heal of your right foot and the foot will pivot on the heal, pointing your toes toward the target. As your body turns, weight will be redistributed onto the ball of the right foot and as your body continues to its final shooting position, weight will once again be on the ball of the front foot. For right handed shooters, the front foot will be the left foot. Always wait until your body has turned down range before mounting the weapon to your shoulder. You will be just as fast as a shooter who cheats and starts to lift the weapon early; the important difference is that you will be safe. And weapon safety is paramount when conducting tactical shooting, in close quarters, surrounded by your team members. Once the weapon is aligned with the target, disengage your safety, squeeze the trigger, fire the round and follow through. Once follow through is complete, release the trigger, safety on, and prepare to do it again.

Right Handed shooter conducting a turn to the right rear

Figure 13.19 Start position for facing to the rear, turning to the right.

Figure 13.20 Head and eyes leading to the target.

Figure 13.21 Stepping with the left foot, make a bold move to allow your body to completely turn towards the threat.

Figure 13.22 Finishing the turn.

Figure 13.23 Drive the weapon to the center of the target.

Left Handed shooter conducting a turn to the right rear

Figure 13.24 Left Handed shooter start position, facing to the rear, turning to the right.

Figure 13.25 Eyes and head leading to the target.

Figure 13.26 Stepping aggressively with the left foot.

Figure 13.27 Finishing the turn, weapon still in the low ready position.

Figure 13.28 Driving the weapon to the center of the target.

Here are some basic guidelines reference; which foot should move.
(Right and Left Handed Shooters)

90 degree left turn = step with right foot

90 degree right turn = step with left

Facing to the rear, left turn = step with right

Facing to the rear, right turn = step with left

All turning techniques should be able to conform to close quarters. By close, I mean extremely close proximity to your team mates. During these missions you will be so close that you may be actually touching or tightly stacked together when you need to conduct these movements. Once you have practiced these drills on an open range and are completely comfortable, practice them near an obstacle that can simulate a team mate. See Figures 13.30A-D. I think you will be pleasantly surprised at how well you can turn and with the speed at which you will be able to place effective fire on a target.

Figure 13.29 Range set up with target stand as team mate.

Figure 13.30A Make your turn as if no one is there, start by leading with your eyes.

Figure 13.30B Shooter starts the turn by rotating his body toward the target as he steps with his left foot.

Figure 13.30C Step forward with the left foot, be aggressive so you are able to make the complete turn.

Figure 13.30D Once the turn is complete, engage the threat target.

Don't get comfortable in a range environment. The world is a 360 degree range, so be prepared to quickly recognize, turn, assess, and engage if necessary. Lead with your eyes, the body will follow.

14. HE WON'T STAND STILL!

Once you have become relatively comfortable with the flat-footed range stance, it is time to move on to shooting on the move. This is where the rubber really meets the road. During Close Quarters engagements, moving will help you to get out of the way of your teammates while working in a hall or transitioning into a room. It also gives the bad guy a harder target to hit. Which targets are harder for you to hit, stationary silhouettes or moving, bobbing targets? The answer is obvious.

To shoot effectively on the move, you must once again apply the fundamentals of marksmanship. To do this, you will have to modify your movement to facilitate quieting the sights enough to effectively suppress the threat that you are engaging.

It will take some time to become confident with shooting on the move. Once you realize it is not as hard as you thought, you will be able to pick up the pace. In this chapter I'll show you several movement techniques that you can practice. These techniques will help you to move naturally and to maintain sufficient accuracy while moving at any angle towards the threat target.

Start Position

Before you begin, you must be in a somewhat aggressive stance with your knees slightly bent. Lower your silhouette by bending your knees. Your knees will act as shock absorbers once the movement begins. If you are stiff legged, your sights will bounce terribly each time your foot strikes the ground.

Figure 14.1 Start position for shooting while moving.

Move Slowly

As you are learning to shoot on the move, do not try to move quickly. Take it slowly in order to build up your confidence.

Start by slowly walking towards the target with an unloaded weapon. Try to keep your sights on the vitals of your target. It will help if you walk as if you were stalking an animal in the woods. Place one foot in front of the other, literally. Try to walk as if you were walking on a balance beam. Do not take large steps. To do so will disrupt your sights with each step that you take. Take small, easy steps, keeping your sights properly aligned and centered on the target.

Figure 14.2A as you start your movement, take small steps.

Figure 14.2E Keep a slight bend in your knees to absorb the shock as your feet hit the ground.

Figure 14.2B Place your feet, heel, edge of your foot, to the toe. Roll forward as though you are trying to stalk an animal.

Figure 14.2F Finish the drill in an aggressive position.

Figure 14.2C If you want to increase your speed take quicker, small steps.

Figure 14.2D Continue the movement, engaging when your sights are on target.

You should also roll your foot from heel, to outside edge of the foot, to the ball of your foot. Once again, move as if you were sneaking up on someone. Your knees should be bent slightly, more than when you are walking normally. This will aid in dampening the shock as your feet hit the ground. Remember, your body must displace any shock or it will show in your sights.

Don't Slide

I may need to further expound on this topic. By shooting on the move, I specifically mean shooting while moving in a somewhat normal manner, not sliding your feet one at a time. Sliding is not an effective way to get you where you need to go and it does not provide a stable shooting base. Sliding will only allow you to shoot as you get to the finish of the slide. If you are able to start shooting while moving naturally, then you will be much more comfortable and will be able to move much quicker. Try not to change the way you move with or without a weapon. Do you normally slide step around the house? Well, maybe, if you are into two stepping on the country dance floor, but this ain't dancing. Walk as normally as you can and shoot.

Using the slide technique is also detrimental to team tactics. It is difficult to move as a team if individuals are slide stepping. This technique, when used for Close Quarters, tends to slow movement of a team once a threat is encountered. This is the most important time to be moving, so don't slow down.

Pause Stepping

Pause stepping or slightly hesitating as you make your shot is also detrimental to truly tactical engagements. If you pause then it will slow down everyone around you. It will also prohibit others to quickly clear the fatal funnel during team tactics. This pause allows the threat to get shots off in your direction, which is clearly not a good thing.

If you have decided to shoot only when a certain foot is coming off the ground, please disregard this technique. Competition shooters can use techniques like this. We, as tactical shooters, can't. If we have a threat to engage, there is no time to wait for a specific foot to leave the ground.

> **Fatal Funnel.** *The Fatal Funnel is the area around a door that a bad guy will focus on. We must clear out of this area as quickly as possible to avoid becoming a target, or worse yet, becoming a speed bump that the rest of our team has to negotiate.*

Moving Quickly

Moving quickly while shooting is a difficult task to master; however, with a little discipline you will not have any problems. Normally in life, if we want to speed up our movement we take bigger steps. To develop good shooting while moving skills, you have to break that particular habit. If you need speed, build it by taking quicker, but not longer, steps. If you take long steps, the impact of each foot will show movement in your sights so use short, smooth steps to gain speed. If you are moving quickly and need to slow to shoot, do so by lowering your center of gravity as well as taking smaller steps. To lower your center of gravity, bend your knees slightly more than you normally would. If you simply start taking smaller steps, your momentum will continue to drive your body. This will create an unbalanced, weight forward feel. Being unbalanced will also take away the control of your body to react to obstacles and will not allow stopping behind cover if necessary. So, as you slow, bend your knees; thus lowering your center of gravity and most importantly, allowing you to stay in control of your movement.

Read Your Sights

As you move, concentrate on your sights. Watch how they jump and then become somewhat steady before bouncing again. Determine that quiet point and try to squeeze the trigger at the correct time. What is an acceptable sight picture for the distance you are shooting? Don't wait on the perfect sight picture. Start your trigger squeeze and accept some wobble or drift. Most shots will end up right where you want them. Also, remember to follow through and read your sights. This is why we practice. If a bad shot breaks, then you will immediately see it and re-engage the target without delay. Remember your primary goal: quickly suppress the threat. Whether that means one shot or five will be determined by what you see in your sight picture as well as how the threat target reacts. I am not saying to get sloppy; just don't take all day. You have a threat, engage it. Follow through, follow through, follow through. Shoot 'til you're happy, and he isn't...

Moving Obliquely

Once you have practiced shooting while moving directly at your targets, change it up and slowly work more and more oblique. Soon you will be surprised that you are shooting 90 degrees off your line of movement. This will take some practice, but you can do it. As you move to the oblique, twist your upper body and don't cross over your legs or feet; let your feet move in a straight line. Some shooters use their High School Football techniques for these crossovers. It does not work well at all, especially at night, or on uneven terrain. If you start to cross over, it will be easy to become tangled or to trip on something you can't see. Stick to moving your feet as close to straight in line as possible. This makes every shooting while moving scenario possible without changing your method drastically. See Figure 14.3A-H and 14.4A-14.4F.

Shooting while moving Left to Right

Figure 14.3A Stay tight on the weapon once it is mounted.

Figure 14.3B Step aggressively but take small steps.

Figure 14.3C Keep your knees bent as you move.

Figure 14.3G Quicker steps, not longer steps.

Figure 14.3D Shoot when your sights are on the target. Don't time your shots.

Figure 14.3H Keep your back somewhat straight, it will be easier when you are wearing all of your gear.

Figure 14.3E Keep moving and shooting.

Shooting while moving from Right to Left

Figure 14.4A Moving Right to left. Mount the weapon and start your movement. You may have to cant the weapon slightly toward your support side.

Figure 14.3F If you are close to the target, you can increase your tempo slightly.

Figure 14.4B When you have mounted the weapon, you may have to slide your support hand slightly farther towards the rear of the weapon.

Figure 14.4C Step across, keep your lower half pointed in the direction of movement.

Figure 14.4D Your upper body will be twisted towards the target.

Figure 14.4E Don't take large steps.

Figure 14.4F If you need more twist, bend your knees. This will allow you to twist more.

Like I said earlier, move as you would normally, or as close to normal as possible. For right-handed shooters, moving left to right will be pretty easy; however, once you start moving against the grain, it will take some work to become comfortable. You will really have to twist your upper body and possibly, slightly cant the rifle toward your non-firing side. Do not stutter step to get the shots off; move smoothly and slowly while shooting. It gets easier and easier as you practice.

As you practice, work on mounting the weapon as you move. As you move in front of a shoot target, let your eyes perceive the threat, present the weapon, disengage the weapons safety, engage the target, follow through, engage the safety, lower the weapon, and repeat. Always engage the safety when you bring the weapon off target. This is a must in tactical environments.

Oblique movement is for team tactics. This will allow you to stay out of your mates field of fire, while placing effective fire on your target. And, as stated earlier, practice the most difficult scenario; it will make real life a lot easier.

When practicing your shooting on the move, always include reloads, malfunction drills, target discrimination drills, transitions to your sidearm, etc. If you practice these drills while moving, it will help you when the bullets start to fly. A problem should never stop your movement. A problem may cause you to stop at the nearest cover, but do not practice stopping every time you have an issue. Fight through the problem.

Zig-Zag Drill

The Zig-Zag Drill is a course I designed to allow the shooter to not only shoot while moving in all directions(except to the rear), but also to incorporate speed of movement.

In the Zig-Zag drill, the shooter starts in the rear corner and moves forward while shooting. The shooter then negotiates a turn and shoots while moving to his or her oblique. Once this is finished, the shooter must place the weapon on safe and quickly move to the opposite corner, negotiate a quick turn to their support side, engage moving forward again, and lastly engage from the oblique, but to the opposite side.

I prefer the Zig-Zag drill over the Box drill. The Zig-Zag drill will make the shooter hustle, negotiate obstacles, open their peripheral vision, and manipulate the weapon's safety. You can also put the shooter on the clock which helps to increase speed while maintaining accuracy.

The Box drill also requires the shooter to move to the rear. I am not a fan of moving away as you shoot because it breeds bad habits. You don't walk backwards on a daily basis, right? Walking backwards is a good way to fall, and fighting from the ground is another can of worms. If you do decide to move to the rear, use the rear security technique below.

Start at either rear corner, moving to the left or right, engaging the targets as you go. You should use more than one target so you can also work on multiple target acquisition. Once you have engaged the targets, moving left or right, and have made it to the other rear corner of the box, start moving forward, engaging the same targets again. Once at the front corner, continue to follow the outer perimeter of the box, engaging as you go. See Figure 14.6A-S, which shows one way of moving around a box drill course. (The Zig-Zag Drill is a modification of the Box Drill Course).

Moving Around the Zig-Zag Drill Course

Figure 14.6A Right handed shooter starts with muzzle touching the cone at the left, rear corner.

Figure 14.6B Once the start signal is given, the shooter starts to move forward.

Figure 14.6C Disengage the safety and begin engaging the first target with three rounds.

Figure 14.6D Once you have finished engaging the target to your front, engage the safety and quickly move to the next cone.

Figure 14.6E As you arrive at the next cone, disengage your safety and prepare to engage the three, center targets.

Figure 14.6F Do not engage the center targets until you have passed the cone.

Figure 14.6G Engage all three targets with two rounds each.

Figure 14.6H After engaging the three center targets. Engage your safety and move quickly to the rear cone.

Figure 14.6I As you move to the rear cone, ensure that you have your weapon pointed in a safe direction.

Figure 14.6M Keep a low profile while moving forward.

Figure 14.6J Keep a low stance as you get to the turn, watch your footing.

Figure 14.6N Once engagement is complete, engage safety and move quickly to the next cone.

Figure 14.6K Aggressively turn the corner as you begin to mount your weapon.

Figure 14.6O Do not engage the targets until you have passed the cone.

Figure 14.6L Disengage your safety and engage the right target with three rounds while moving.

Figure 14.6P Twist your body. Keep your feet moving towards the next cone.

Figure 14.6Q Engage each target with three rounds.

Figure 14.6S Once finished with the drill, ensure that you engage your safety.

Figure 14.6R Bending your knees will help you to twist enough to engage the targets.

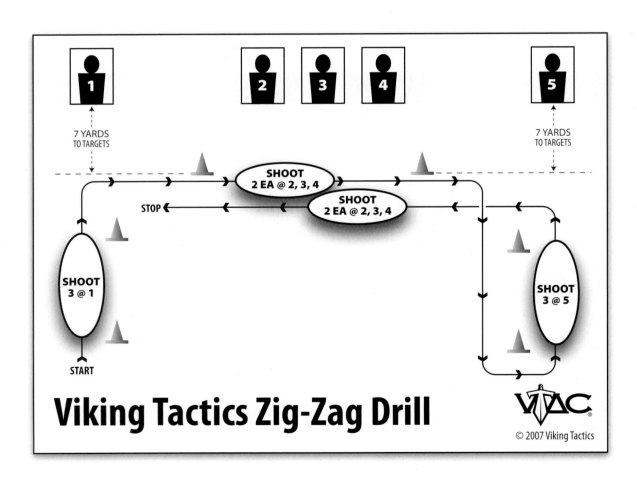

Rear Security Technique

Instead of walking backwards, I walk using the rear security technique. This allows you to quickly stop your movement and engage if you have to shoot to the rear of your formation. This movement is conducted as if you were walking away from the target, looking over your non firing shoulder. I keep my head on a swivel, looking forward, then quickly behind me again. Once I have a threat to engage, I keep my eyes on the target and quickly make a turn to the left and engage the threat. This will be the opposite for left handed shooters. This stance is not ideal for pulling security, but it works well for quickly moving and maintaining some form of security.

Figure 14.7B As you start to move, turn your head to look over your support side shoulder.

Figure 14.7C If you are stepping forward instead of to the rear, you won't be as likely to fall if you trip.

Figure 14.7A If you are in a static position, you can have your body twisted toward the unknown.

Figure 14.7D Keep your head on a swivel.

Figure 14.7E If a threat appears.

Figure 14.7G As you square to the threat, Drive your weapon to the threat.

Figure 14.7F Lead with your eyes to the threat. Start stepping towards the threat.

Figure 14.7H You will see that this technique is extremely fast and you will be very sure footed.

15. TARGET DISCRIMINATION

Right versus wrong, good versus evil, or, as I have said before, evil versus more evil; however you want to analyze the situation, we need to be better than the bad men we face. In addition to needing to be better marksman, quicker thinkers, stronger, faster, etc., we also need to discriminate between which targets we engage. If we were to indiscriminately engage anything that moved, we are no better than the wretched individuals whom we are hunting. If you are a Law Officer, Soldier, or protecting your home, you must analyze shoot and no shoot situations as quickly as possible. If you do not learn to discriminate between targets, then you won't last long in the gun fighting world. It is extremely important to be quick, but it is even more important to engage the correct threat target and not an innocent bystander.

When training for target discrimination, we need to first decide what your chosen profession will allow. Can you shoot an armed individual as soon as you perceive there is a threat, or do you have to give pre designated, verbal commands before you can engage? It is up to your Department or Service to decide your rules of engagement. I will give you the discrimination tools, but you must decide when to use them.

In previous chapters I have discussed weapons safety, target to target acquisition, and leading with your eyes. These small tasks, when lumped together, are paramount to success in a shoot/no shoot situation.

If you disengage your safety before you have a target to engage, you are wrong.

If you do not lead with your eyes, you are wrong.

As your eyes lead into the target, you will be able to momentarily scan the target's hands, as well as to determine his demeanor. If you determine that there is a reason to engage this target, then you will have time to place your sights where needed, disengage the safety, and destroy this individual. When you are on the range, practice as if you are engaging threat or non threat targets. The only way to get better at quickly assessing the status of targets presented to you is to practice.

When you begin practicing target discrimination, start small. Painting pistol outlines as well as hands on all of your targets is a good starting point. If the target has hands with no weapon, then it is a no shoot. If there are hands with a weapon, engage. See Figure 15.1 for targets with hands and guns painted on them. As you become more confident, have your range buddy set up your targets for you. Do not cheat yourself by peaking. Use a vision barrier or simply cover or close your eyes until it is time to start the drill. Once the signal is given, quickly assess and engage the correct targets. Ensure your set up is safe and that your partner is on the same wave length as to what your particular range will allow.

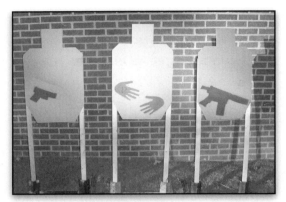

Figure 15.1 Targets with hands and guns painted on them

Once you have built confidence, you may take the training to the next level by using cartoon shoot/no shoot targets. See Figure 15.2. I recommend cartoons initially since they usually are easier to quickly discriminate. Cartoon targets are targets that are not real people or pictures of real people. They are cartoons. Sounds slightly silly, but it works.

Figure 15.2 Discrimination Targets

Once you are confident with cartoon targets, acquire realistic shoot/no shoot targets. These targets will look like real people and should allow you to place weapons, badges, cameras, beer cans, whatever you decide to use for confusion, on the targets. These targets are available from most Law Enforcement stores or online. When you purchase targets, also buy the hands and weapons that match your particular targets. If you do not have the ability to change the targets, the drill will be too easy and you will gain nothing from the training.

When you use discrimination targets, you will still not get the full benefit of engaging real people. Paper targets are not going to bob and weave as erratically as a human target. You will also not receive changes in demeanor from these targets. Therefore, once you have trained to proficiency, you will need to practice discrimination exercises with real people using paint marking weapons or soft air training weapons. These weapons are one of the best tools available today for training the professional to discriminate between targets. Not only does it help you to assess the status of the individual being confronted, but you are also able to read body language or demeanor.

These target discrimination skill sets will degrade quickly if you don't practice. It is important to make shoot/no shoot drills a standard staple in your team and individual training. There is a fine line between mission success and mission failure. Mission success requires every member of your team to have the ability to effectively and proficiently discriminate between targets.

15B. TRANSITIONING TO THE SIDEARM

In Chapter 5 we discussed malfunction clearances. It is extremely important for malfunction clearance drills to become second nature to soldiers. These men normally do not carry a sidearm, or as discussed in Chapter 5, if they do wear a sidearm, the engagement distances may be out of range for their pistol.

For those operating in a little different environment, such as a close quarters situation, the choices may be slightly different. Tactical shooters who also deploy with a sidearm have an advantage at close range if their primary weapon happens to malfunction. These shooters may also come into contact with someone who decides to test their weapon retention skills. Whatever the situation may be, I will discuss techniques for quickly getting your backup weapon system into use.

The first important thing to remember when conducting transitions is the fact that you are training for a gunfight. Treat the training as such. If you have a stoppage, do not try to assess the malfunction in a close encounter. Immediately start the transition process. It does not matter what has caused the malfunction, the fact of the matter is that you have no bullets exiting the barrel of your primary weapon, so get the back up out and into use as quickly as possible.

You may get a click of the hammer falling without a large bang; you may actually feel the weapon go dry, or run out of ammunition, and still have a threat to deal with; or you may just have a dead trigger, a trigger that you squeeze and nothing happens. If you have any of these indicators, start the transition process.

Using the Safety

Inexperienced shooters will tell you that the safety must be engaged before you transition to the sidearm. This is simply asinine. Why do we transition to the sidearm? Either you have had a malfunction that will not allow the weapon to fire, or you are out of ammunition. Either way, the weapon is in a relatively safe condition. You, however, are not. You have a threat that must be quickly eliminated. If the weapon has had the hammer drop to the forward position, the weapon should not go on safe anyway. If it does, you need to take the weapon to a competent gunsmith and get the situation rectified. If your weapon is out of ammunition and the bolt locks to the rear, you will be able to engage the safety, but why?

My point is, if your primary weapon malfunctions or is out of ammo, it is important to get to the task at hand and to not waste time. If the drills you are using to practice this skill set do not realistically represent what will happen in an actual malfunction or dry weapon situation, then change the drill. Later in this chapter I will give you several drills to accomplish this training in a very realistic and safe manner.

Transitioning

Now that the safety manipulation issue has been covered, let us get on with the act of transitioning.

After you receive the indicator that the weapon is no longer of any use in this gunfight, what do we do? I have seen techniques where the shooter simply lets go with both hands and goes for his pistol. This technique is no quicker than what I will teach, and it comes with the extra advantage that you won't burn yourself with a hot weapon or hit the family jewels when the weapon drops between your legs. Practicing transitions should not be painful.

After deciding to transition, immediately release the weapon with your firing hand. To be more specific, release the weapon with the hand that will draw the pistol, since there are those out there who shoot one weapon right and the other left, or vise versa. Once the drawing hand is free it must go as quickly as possible to the sidearm. Snap the hand to the weapon then slow down to establish a good hand position for the draw. As you drop the hand to the pistol, use the non firing hand to slowly guide the weapon to your side. It doesn't matter where you end up with the rifle, because it is, in effect, just a club at this time anyway. This is of course contingent on you having a decent tactical sling that allows the weapon to hang in a muzzle down configuration.

Once the weapon has been lowered, and you have released the weapon with your non firing hand, simply grasp the pistol with the non firing hand and present it to the threat. Your transition draw stroke should be as near to your normal draw stroke as possible. It shouldn't require any whacky movements to get the weapon on target.

If you are in an aggressive stance with the rifle, you may need to take a slight step forward with the firing side foot as you draw. This small step should happen as you draw, not after the draw. The small step should not impede getting the weapon to quickly bear on the threat.

Transitioning to Sidearm

Figure 15B.1A Transitioning to the Sidearm. Sling adjusted to keep weapon from dropping too low.

Figure 15B.1E Finish lowering the rifle and move support hand directly to meet the pistol.

Figure 15B.1B As soon as you sense the need for a transition, the weapon will start to drop.

Figure 15B.1F Once the proper grip is established start driving the weapon to the target.

Figure 15B.1C As the rifle is lowered your strong hand immediately goes to the pistol.

Figure 15B.1G Push the pistol directly to the target, do not porpoise, or dip the muzzle.

Figure 15B.1D Continue lowering the rifle as you attain a firing grip on the pistol.

Figure 15B.1H Finish by putting effective fire on the threat. Notice, VTAC sling keeps weapon out of the shooters way.

Transitioning from Support side to sidearm

Figure 15B.2A Transitioning from Support side to pistol.

Figure 15B.2B Maintain control of the weapon, immediately move strong hand to pistol grip.

Figure 15B.2C Ensure grip is established close to the body for good weapons retention.

Figure 15B.2D Drive the weapon to the target. Not sling will be hanging just around your neck, as soon as possible, re sling the weapon.

Sling Adjustment

If you have an adjustable sling, this is the time to make sure the sling is adjusted correctly. The sling should only be just long enough for you to be able to fire the weapon from your normal stance. You won't need a lot of extra. This will keep the weapon slung tight as you conduct the transition.

If you are using a three point sling, then you may have issues with the weapon hanging too low after the transition. You should only have to guide the weapon down a short distance with the non firing hand. If your sling adjustment does not facilitate this, change slings, or if possible, adjust the sling.

The weapon may be hard broke, so as I have said before, train for the worst. Can you continue your mission with just that pistol? Will your slung weapon impede movement? If so, adjust the sling until it can actually be used for this purpose.

Weapon Transition Practice

As stated earlier, the drills we conduct to enhance our transition capabilities should be as realistic as possible. The more realistic the drill is, the safer this drill will be. If you are conducting transitions with your weapon still hot or loaded, you are wrong.

I normally begin my transition practice without even using a rifle. Once I am on the line with the pistol holstered, I conduct my draws as though I am transitioning from the rifle to the pistol. This will help you with the transition. If you conduct these half dry drills, the actual transition will become second nature. All of your drills should replicate what will be conducted in a real life situation. There are certain drills we use that are strictly to build confidence with our weapons manipulation skills. Reflexive gun handling skills are necessary because there will not be time for a do over in a combat situation. This being said, if the drill falls apart, finish it anyway. If you performed poorly, do it again. Crazy things will happen when Murphy shows up at the party, so train hard for his arrival.

Weapons Transition Drills

<u>Dry Fire Drill.</u> The dry fire drill with the transition gives you the same training as live fire, just with less of a bang. With your rifle and pistol clear, conduct an up drill on the target with the rifle. Once you have dropped the hammer to a click, quickly transition to the pistol and dry fire it as well. I always start my transition training with a few such dry runs as I just described.

<u>One Shot One Click.</u> The first live fire drill we will talk through is the one shot, one click drill. You can set up this drill quickly so you don't waste time loading several magazines. It requires one 30 round magazine, the total number of rounds in the 30 rounder is up to you. Once on the line, make sure your pistol is loaded, or unloaded if you wish to conduct dry fire with the pistol. Dry firing the pistol during transition training is a good way to start. Using the building block approach will normally build better skills and create confidence in new shooters.

After you have decided on the status of the pistol, place it in the holster and fasten your thumb break. You must train as though this is a real situation, and most folks have their thumb break engaged when using the rifle in an actual confrontation.

Make sure the rifle's safety is on at this time.

Load the rifle with one round, using the loading technique that we described earlier in this book. This will also help you to build muscle memory for loading the rifle correctly. The only change to the sequence is that you won't put the magazine back in the rifle after you check the magazine to see if a round was chambered.

You are now ready to conduct the drill. Mount the rifle using the aggressive shooters stance. Don't change your stance just for this drill. Every time we conduct a drill we want it to be as close to our normal set up as possible. Once you have mounted the rifle and aligned your sights on the target, squeeze the trigger. After the round has fired, follow through and squeeze the trigger a second time. You will now get a click, which of course signals us to conduct our transition.

Quickly lower the weapon while drawing the pistol. Complete the drill by engaging the chosen target with your pistol.

This drill is relatively simple, but it works very well during transition training. Reload the rifle and repeat as necessary. Always ensure that you place the rifle back on safe before reloading to repeat. This is a common oversight with this drill.

<u>One Shot To Bolt Lock</u>. The one shot to bolt lock drill is similar to the first drill (one shot, one click) we conducted, with one exception. We have to start with a magazine loaded with only one round. Load the rifle as you did with the one shot, one click drill. Instead of removing the magazine, leave the empty magazine in place. This will allow the weapons bolt to lock back after the weapon has run out of ammunition.

Conduct the drill in the same manner as the previous drill. The only change will be that the bolt will lock back after the first round is fired; or it should, if the weapon is working correctly.

This drill accomplishes two goals. First, you are able to conduct your transition training. Second, you will start to learn the feeling of the bolt locking to the rear. Learning what this feels like can save time in a real world situation and may actually save your life. You may be excited and not reload as needed. Being able to sense how the rifle feels as the bolt locks to the rear gives you a jump on either the transition or a combat speed reload.

Dummy Rounds

With the previous drills you have a pretty good idea what is about to happen. When you are initially practicing these drills you will become more comfortable with what you need to accomplish a good transition.

Once you have become comfortable, then it is time for a little more unknown. I like to use dummy rounds to surprise the shooter. Placing the dummy rounds somewhere inside the magazine, throughout the load will cause the shooter to be somewhat surprised. Normally the shooter knows that there are dummy rounds, however the unknown is when they will get to that round in the magazine.

These dummy rounds work well for conducting malfunction clearances, but they work just as well when working through transitions.

See Figure 5.41 Orange Dummy Rounds.

Figure 5.41 Dummy Rounds

<u>5-7-9 Drills.</u> To make normal shooting drills interesting, have your shooting buddy load three to four magazines for you with five, seven, or nine rounds. This will help

to keep you on your toes. Most drills are conducted with even numbers of rounds; loading an odd number will make the transition come somewhere in the drill, not at the end of the drill. Downloading the magazines to 5-7-9 will also allow you to conduct the transitions several times without burning through a lot of ammunition.

Do not hesitate when you are conducting transitions. If you are in such a position that transitioning to the sidearm makes sense, that little switch in your head should click to let your brain know that transitioning is an option. Once this switch clicks, be prepared to quickly transition. Do not slow down to assess the status of your weapon; this isn't necessary. What is necessary is quickly and decisively eliminating the threat that you are confronted with. Once you have become comfortable with the transition practice, push your speed. It is always better to be quick during your practice sessions. Training at 100% will allow you to slow down slightly to 90% during a real world encounter, but still to be exponentially faster than the bad guy you are fighting.

16. TERMINAL BALLISTICS AND OVER-PENETRATION

Terminal Ballistics, for our purpose, is how the bullet performs once it strikes a human body; or, as in the shooting industry test protocol, it is what happens when it hits Ordnance Gelatin. Normally the desired effect is as follows: The bullet strikes the body, penetrates slightly, the bullet starts to become deformed, or mushrooms. As the bullet continues through the body it creates a temporary wound cavity as well as a permanent wound channel. While the bullet is coming to a stop or passing through the body, it transfers its kinetic energy from the bullet to the body. This kinetic energy should further help to disrupt the internal organs. The end result is the threat being incapacitated.

So what really happens when the bullet strikes? Better said, what is the preferred performance when a bullet strikes its mark? When you get ready to select a bullet for something other than target practice, you should follow a few basic guidelines.

How accurate is the bullet, or better said will the bullet perform to your standards of accuracy? If the answer is yes, continue. If the answer is no, keep looking.

The bullet should be designed to penetrate easily to vital organ depth. If you select a varmint bullet for use against two legged, pop up, shoot back targets, you will not attain the minimum amount of penetration to reach important vital organs. You cannot always plan to strike straight into the chest cavity, you may need to penetrate an outstretched arm and then into the chest cavity, or engage the target from a sharp oblique through a thick pectoral muscle. Either way, a round that only penetrates 3-4" will not cut it. A good rule of thumb is a minimum of 12" of penetration. Even if you chose a bullet that does not mushroom to the desired circumference, at least you will have the penetration depth to hit something important. How many mobsters have been killed with an ice pick?

Expansion is also a key feature to investigate. Obviously bigger is better, within reason. Ideally your bullet should expand very near the surface of the target, continue to crush and tear organs as it travels to a minimum depth of 12", coming to rest just inside the opposite side of the body or exiting without enough gas left to hurt innocent bystanders.

This sounds relatively simple, especially in this day and age of the internet, cloning, and the orange mocha frappuccino. There are, however, several other factors that come into play. How far away were you from the target? Did the bullet bleed off enough velocity causing the bullet to under expand and over penetrate? What happens if the bullet strikes a bone?

How will your bullet perform against barrier material? Can you shoot through a car windshield, from the outside? How about from the inside? What happens if you shoot through sheet rock? These are all questions that need to be considered.

Bonded Bullets

Bonded bullets are designed to aid in the retained weight of our ballistic projectile; however, if we have retained weight, then we also gain penetration. This is good to a certain point. It is up to you and your Department or Unit to determine what your minimums and maximums should be. I feel a bonded bullet or a bullet designed with the characteristics of M855, more commonly referred to as green tip, have most of the characteristics I prefer. They work well through glass and through plywood, as well as several other types of material; and once they strike the flesh of a human, they certainly perform as advertised. This, of course, is dependant on velocity. Now back to a previous question, at what velocity does your bullets performance degrade? Have you shortened your barrel so much to make the weapon easier to handle in confined spaces, leaving the velocity part of the equation suffering?

Over Penetration

There are several myths you hear about over penetration on the range. "Don't shoot 5.56 in a house or CQB situation. It will over penetrate if you miss your target. Even if you hit your target, it will over penetrate." Of course my favorite is, "we need 45 ACP sub-guns so we don't have over penetration problems with our ammunition". This is a farce, on average, 45 ACP will penetrate much farther than 5.56. This is especially true when you are talking about flesh or gelatin shots. 230 grain FMJ (full metal jacket) ammunition is one of the deepest penetrating types of ammunition you could possibly carry. Try up to 30 inches of penetration. Compare this to most 5.56 ammunition in FMJ configuration and you will see about half the depth of penetration. Now, obviously the addition of an expanding or hollow point bullet helps to reduce the problem of over penetration, but it is still important to know how far they will penetrate if the hollow point somehow becomes ineffective. For example, after firing through sheetrock or heavy clothing the hollow point is completely plugged with foreign debris. This may cause poor performance, which may mean over penetration.

The age old question is this:

Can a 5.56mm bullet instantly incapacitate a human? The answer is simply **yes**.

The next question to answer is this:

Can a 5.56mm bullet hit solidly in a human, yet the threat continue to fight? Once again the simple answer is **yes**.

This may not make a lot of sense to some of you, but we cannot find a magic round that will incapacitate a threat on impact with any part of the body. It is just not physiologically possible to have a constant when it comes to terminal ballistics. Every person is different. Every bullet performs slightly different in the media that is used. Did you strike a rib? Was the shot from an oblique?

Bigger is obviously better when it comes to these questions, but at what point do we give up reliability and magazine capacity. What about the shootability of the rifle?

If you are an animal hunter, you have seen deer or elk that have been hit solidly with large caliber weapons, yet continue for some distance. Human hunters have the same issues to deal with. Once you accept this fact, life is so much easier. To remedy this situation, we must continue to engage the threat until it is neutralized or is no longer available to be shot. Don't expect one shot or even two to get the job done. This goes for those shooting .308 and larger calibers as well.

What are we trying to hit? I like to picture the target as a piece of re-bar with a couple of tennis balls skewered on the re-bar. The two tennis balls represent the brain and the heart. The re-bar represents the spinal cord. Is this a difficult target to hit? Of course, but it is realistic.

The 5.56mm carbine has been used successfully for many years to eliminate threats throughout the world. There are some who have had a bad experience with 5.56mm. The only answer I have for their displeasure is to work on accuracy. Real people are not easy to incapacitate and paper targets are much easier. If you have ever witnessed first hand the damage that the 5.56mm projectile will cause to the human body, you will become a believer. Don't believe everything you read in a gun magazine. Their obvious goal is to have something new to write about. They don't have the shooters best interests in mind. If you really want to hear the truth, talk to someone who has been there...

17. NIGHT SHOOTING
Bad things happen at Night!!

Many shooters overlook night shooting because they think it is too inconvenient, too difficult, or just not as fun as shooting in the daylight. I will have to let you in on a little secret. Bad things happen at night! Most engagements take place during hours of limited visibility. So why not train in these hours of limited visibility? Night shooting is a must for the tactical shooter. Training to build confidence during the day is important, but it is just a dress rehearsal for what needs to happen at night.

What works at night?

There are several ways to shoot well at night. We have all kinds of special tools that we can choose from. We have white lights, visible lasers, IR lasers, tritium sights, and thermal imagers. Probably the easiest item for any of us to get a hold of to increase our night fighting capability is the white light. The white light doesn't require any special permits or Night Vision Goggles to achieve decent results. The white light can usually be mounted without a lot of special tools, and it allows you to use your existing day sights, not requiring night zeroing.

Light Selection

Which light is the best to mount on your rifle? The best light is the light that you have. If you have a mag light, it is better than nothing. I prefer to use a Surefire 6P or G2 Nitrolon. These two incandescent lights will fit into most mounts that are available; they are also relatively inexpensive and very durable. You do not need a shock isolated bezel when the light is mounted on your M4 or AR type rifle. Surefire has also started to offer replacement LED bulbs for the G2 and the 6P.

If you really want to be high speed you can purchase a LED type light system. These lights are extremely reliable; however, they do have their downsides. LED lights make target discrimination slightly more difficult. It is a little harder to see what is going on near the periphery of the light. This is not a huge issue but it does have an impact on how quickly you can decide what you are actually seeing. The up side to this type of light is that the bulbs are virtually indestructible. I believe we will see these lights become better and better in the next few years. The last disadvantage is the lack of transmission of light through an IR cover. If you are using an LED light, you will not be able to illuminate with your IR light cover on. This comes in handy when using NVG's (Night Vision Goggles). If you don't use NVG's, then it isn't an issue at all.

Light Placement

The ideal place to have your light attached would be at the front of your barrel, extending beyond your flash suppressor. Of course, this is not possible in most cases. Having your light extend beyond the muzzle also causes damage to your light when you place the weapon muzzle down in a helicopter or in a vehicle. Another reason not to place the light this far forward is the muzzle blast. If you have the light even with the muzzle, it will be very hard on the light bulb as well as the carbon will build up on the lens after serious use.

I prefer to get the light out as far as it can be and still be comfortable. Like I said before, ideally you would have the light beyond the end of your barrel. The reason for this is the shadow that will appear if the light is too far to the rear, a common issue with shooters who have suppressors on their rifles. They are good to go until they attach a sound suppressor and do not change the placement of the light. If you do not change the light placement and the light is attached to the six o'clock position, then you will end up with a shadow that covers the head of your target. This large shadow will cover key parts of your target, namely the eyes. See Figure 17.1 Shadow from light on target head.

If you have ever seen the movie Tombstone, you will understand what can happen if you can see the bad guys eyes. In one scene a fellow actually winks at the cowboy before the gunfight starts. He smiled then winked. That action set them all off. This is just a movie, however if you are planning to see a bad guys demeanor, you will normally see the angst in his eyes if he is getting ready to make a move. Thus, it is important to be able to have a light that does not put a shadow on the target's eyes. You must also remember that we are using the light as a tool to give us an advantage. We shine the light into the bad guy's eyes to have a startling or controlling effect. If there is a shadow over his eyes, he will be able to see you too. Not a good plan if you are looking for every advantage possible.

Figure 17.1A Shadow from barrel on the target head.

Figure 17.1B No shadow on the target, light mounted farther forward.

Some shooters pull the light as far to the rear as they possibly can. This does help with getting the center of gravity a little more in your favor, but the disadvantages out weigh this particular reward. The most important issue to the tactical shooter pertaining to light placement is the fact that you are giving the bad guy a target to shoot at. I would much rather have this target as far from my head as possible. It is bad enough that the threat can engage you straight from your front, don't give them a target from your flanks.

Using the light

The light should only be used when necessary. This includes any time you bring your weapon to bear on a threat target, day or night. We must gain every advantage we possibly can. As I stated earlier, use the light as a distraction to the threat, he should be confused by the bright light. He will also not be able to see what is going on as long as the light is directly in his eyes.

Light Discipline

So when is a bad time to use a light? When you are sneaking up to a building to serve a warrant, having a white light AD (accidental discharge) is unacceptable. If you are the second man in the door or in a hallway you should not be back lighting your partner. Look where his light shines if you need to see where to step. It takes a little practice, but you will have no problem once you try it a few times. Also watch your light discipline when operating around windows or near closed doors. Any indicator to the bad guy gives him the go ahead to shoot. He doesn't have to see his target, he can shoot through curtains, through doors, and of course through walls. Be careful how much light you use.

I prefer to use the push button on the back of the light, rather than a tape switch. Tape switches are trouble. They are not as durable as the button on the back of the light. They also will cause more white light AD's too. If you start to fall from a ladder or stumble, you will naturally grab whatever is near to keep from falling. If you have a tape switch it can get pressed at the worst possible moment. They also have a tendency to turn on in your weapon case or gear bag. Not too cool to pull out your rifle and the bulb is burnt or the batteries are dead.

Figure 17.2 Tape Switch

The push button on the back of the light can be set up in such a manner to allow left and right hand access as well. This is important for the tactical shooter. You must be able to apply light to the target, left or right handed. See Figure 17.3.

Figure 17.3A Light Activation with Left Hand.

Figure 17.3B Light Activation with Right Hand.

Do not leave your light on. You must be able to turn your light off quickly. I prefer the push button set up for exactly this purpose. If I let go the light is off; if I really want to, I can depress the switch to turn the light on constantly, but I won't do this until the chance of contact is over. If you have your light constantly on, chances are that you will be the guy who gets shot. Now everyone around you will be lit up with your constant light. Not a good idea.

Visible Laser

Visible lasers are becoming more popular. I was never a fan of the visible laser until I had a chance to use them in real confrontations. Armed individuals didn't escalate the situation after seeing the lasers bouncing around on their chests. They simply raised their hands allowing us to move closer and handcuff the individuals. They also have worked well to alarm suspected terrorists on the sides of roads. When these sketchy insurgents saw the laser moving on the ground toward them they split. Were they about to detonate an IED (improvised explosive device)? I don't know, but they took off running and we didn't get blown up. Laser discipline is necessary just like with the white light. If you can see it, so can the bad guy. Only use the light when it is necessary.

The visible laser can also be used for booby trap detection. If the device is trip wire activated you will be able to see the wire with the visible laser. This works better if you have two individuals working together. One shining the light, the other looking for the splash, or bright spot that will develop as the laser crosses the wire. It isn't fast, but neither is evacuating someone who has been injured.

Laser Rule of Thumb

There have been several innovations with regards to lasers in the last several years. Even with these innovations, we still must use a swag (scientific wild ass guess) to get our lasers zeroed.

If you hear a manufacturer tell you that they have a way to zero the laser for several different distances, they might just be pulling your leg.

Here are a couple of good rules of thumb when it comes to laser attachment and zeroing. First the laser should never be mounted below the barrel. If the laser is mounted below the bore, then you will be zeroed for the one yard line but way off at the other distances. The laser shines in a perfectly straight line, in layman terms anyway. As the bullet flies through the air it has a curve, this curve increases as the bullets flight speed decreases. If you are still with me continue, if not please go back and read this section again. It is important that you understand what is happening with the laser in relationship to your bullets trajectory.

The next rule of thumb: Never get an exact zero with the laser. If you are dead on at one yard line, you will be off at the next yard line. If you had to hold on the target with your normal sights, but had to add left or right windage as the distance increased would you want to have a zero like that? I'm guessing probably not. This is where Parallel zeros come in.

Parallel Zero

To get a useable zero with the laser on the carbine for extended distances we must use a parallel zero. A parallel zero may be hard to understand, but once you have played with it for awhile, you will see all the advantages of this zero.

A parallel zero allows you to hold your normal elevations, the same as your red dot sight, optical sight, or iron sights. For example with the parallel zero at 50 yards, your hold for 100 yards will be the same with the laser as it is with the daylight sighting system.

To attain a parallel zero we only need to know the horizontal distance from the center of the bore (or bore axis as it is called). Once you know this distance, you are ready to begin the zero process.

Ideally you will set the zero for the same elevation as your day sight. So if you have a 50 Meter zero during the day, try to get the same zero at night with the laser. The only change will be that we will want the bullets to strike left or right the same horizontal distance that you previously measured. If the laser is mounted on the right side of the rifle, your bullets will need to strike the correct distance to the left of the bulls eye. When zeroing make sure you are holding the laser in the center of the bulls eye. If you are trying to hold off during the zero process you will never get a solid zero.

If your laser is mounted to the left, your bullets will strike to the right of the bulls eye.

If the zero is being conducted with IR lasers and NVG's, use a small piece of Glint tape (the reflective tape that is on all the ACU shoulder pockets). If you are shooting a visible laser, you can use Glint tape or a small piece of reflective tape. Either way, you want an indicator as to when the laser is in the exact center of the target. When using Glint and NVG's you only need a piece the size of your smallest finger nail. Any larger and you will not be able to get an accurate group. Use a piece of tape or small amount of spray glue to hold the glint in place.

Every laser is slightly different. Ensure that you read the instructions before you go to the range at night. You will need to know which way the bullet impact will move when you turn the knobs. Don't try to over think the process. Normally lasers are set up with small arrows that point in the direction that the bullet will move if you turn the knob toward that arrow. Like I said earlier, it is important to read these instructions during the daylight. If there is more than one setting on your laser, memorize which one is which so that you will be able to adjust the laser to the proper setting without using a light. You will need to be completely comfortable with this piece of kit if its operation is to become second nature.

Once you are on the range, remember to use laser discipline. Only turn on the laser when you are shooting. Also, do not shine your laser at people on the range. One, it is bad etiquette. Two, most good lasers are not eye safe. Three, you will look funny with that rifle barrel wrapped around your neck.

Quick Zero Confirmation

It is always best to confirm zero with live fire at a known distance. If this isn't possible, then you do have other options. One option is to check alignment with your scopes reticle. This type of confirmation is only possible if you have a reliable day zero. If not, you won't be able to accomplish anything with this technique.

You should always check the alignment during your zero procedures. This will help you to quickly check after rough weapon handling to be certain that the laser has not been bumped from the original zero.

Complete the zero process as stated above, and then make a check to see how the laser is aligned with the scope. Remember this picture. At a later date glance at the reticle to see if it is, in fact, aligned as you remember. If not, you can make small corrections. If you are not confident in your ability to make these adjustments, try to find a range as soon as possible to confirm. Even if you can't conduct the zero now, make use of live fire confirmation at the next opportunity.

NVG Shooting

Shooting while wearing NVG's is an awesome advantage. Those that have goggles know the huge advantage that this capability can give the tactical shooter. However there are several poor techniques that come from being lazy while shooting with the NVG's and an IR laser. As I stated earlier, using the parallel zero will help you to engage at changing distances relatively easy. But what else can we do to make shooting the laser easier? One of the first ways to increase your night fighting advantage is to shoot from your normal shooting positions. Don't change your stance or position just because you are now using a laser. If you can't get a good stock to cheek weld, then set your NVG on top of your day optic. This not only allows you to tightly mount the weapon, but it also allows for slight adjustments of your NVG's without using your hands. That is, adjustments in which direction your NVG's are pointed. If you plan to shoot while looking through your scope then this technique will not work.

IR Scope Settings

If you happen to have a scope such as the EO Tech Holosight, or the Aimpoint with the IR settings for the reticle be careful how dependant you become on this tool. I only use this reticle setting when attempting to zero my laser. Or when I need to quickly confirm that my laser is still zeroed. You will be extremely slow trying to align your NVG's with the reticle of these scopes. It can be done and practice will help, however you will never be as quick as with an IR laser. There are also systems that allow you to mount your NVG to the rear of the scope. This is obviously a disadvantage for several reasons. One reason is the fact that you will only have night vision when looking through the scope, which makes scanning an area or moving difficult. I much prefer to have my night vision on my helmet. This setup is also extremely slow when trying to engage moving targets, you will have to search while looking through the scope. It is much quicker to scan with the NVG's attached to your helmet, acquire the target, turn on the IR laser, and engage the target.

Sniper systems are a completely different animal, but this is about the use of the carbine, so do not confuse the two.

The biggest disadvantage when using the IR reticle in the scope instead of the laser is the fact that you cannot quickly transition to white light while using this setting. If you are moving with the scope set to an IR setting and you make contact that may cause you to transition to white light, you will not have a reticle to shoot with. This situation will require you to re adjust your scope to a normal setting; not good in a quick engagement situation. One remedy for this situation is to use your BUIS (back up iron sights) for these engagements. The moral of the story here is that you must think through the situation before the time arises, not after.

IR laser/spotlight

If you have an IR laser that is also equipped with a spot light or flood light, it will give you a huge advantage. But to be advantageous, the flood light must be adjusted correctly. Once your laser is zeroed, adjust the flood to the desired size. Once you have the flood large enough to give you an advantage, adjust the flood in the right direction to center the laser dot. Make sure you are adjusting the flood to the laser and not the other way around. Once you are in a situation to use the flood, play with the width of the circle to see what works best for you. As you conduct a mission don't be afraid to change the setting of the flood to optimize for the specific scenario. The setting you would use for the open desert is different than that which would be used for urban operations.

Using the IR flood will allow you to conduct quick searches of dark alley ways. The flood will also help you to scan large buildings and illuminate through windows and back in the shadows of rooms. This is one advantage that IR has over thermal imagers. Thermal imagers will not allow you to look through windows; this is a definite disadvantage in a tactical situation.

Summary

If you have never been lucky enough to train at night, get out there and find a range that will allow for training during hours of limited visibility. This training is priceless in the tactical community. Truly tactical shooters have realized the importance and train in the dark every chance they get. Always train for changing situations. Are you going to be operating around intermittent street lights? Around on coming cars? During dusk or dawn? All situations have their own idiosyncrasies, learn from each situation.

18. MAINTAINING YOUR WEAPON SYSTEM

Maintenance is often overlooked by shooters and instructors alike. It doesn't seem to become a popular topic until you are having trouble with your rifle or you see how well someone else is shooting. Increased reliability and accuracy are a 'no brainer' for the tactical shooter, and a little tender loving care of your gear may have huge benefits in the long run. This is a shooting instruction book; however, I think I would be remiss if I didn't cover maintenance of your weapon. This is simply a simple guideline. If you have a specific way of cleaning and maintaining your rifle, drive on. It might not hurt to read through the following short paragraphs to see if any of these ideas can be added to your system.

For those of you who know me, you have heard me joke about not cleaning my guns. I shoot them until they are dirty, and then I sell them. This briefs well but isn't very close to reality. I definitely do not waste a lot of time cleaning my rifle; I do, however, clean the parts that matter. Having every little bit of carbon removed from the bolt isn't absolutely necessary; having a well lubed action and a clean barrel is much more necessary than the carbon removal.

Basic Tool Kit

Before we discuss the cleaning process I would like to lay out a basic tool kit that will help you to keep your rifle running. The only necessities on the list are a Leatherman tool and a good cleaning rod with bore brushes. Everything else just helps to make the job easier.

Leatherman tool: I prefer the Leatherman Wave. This tool allows access to the knife blade without opening the handles. I also prefer to have a tool with a needle nose rather than flat nose pliers. This thin nose allows for pin removal and can also be used to assist in clearing a stuck case from the chamber. The Wave also has several other tools that are handy when using the tool in the tactical arena. There is a file, saw blade, several screwdrivers (one of which is large enough to tighten the butt stock screws of an A2 Rifle), and, of course, the coveted can opener.

Cleaning Rod: I have several cleaning rods. The rod I prefer for cleaning my AR is a stainless steel Dewey rod. These rods are well made and have a ball bearing handle that rotates easily. I also use a GI cleaning rod when necessary. The GI cleaning rod can serve two purposes; one being the cleaning of your rifle, the other is for clearing stuck cases. There is also an ingenious product called the Bore Snake. These work well to give your weapon a quick barrel swab, but they don't work quite as well as a conventional cleaning rod.

Figure 18.1 Dewey Cleaning Rod

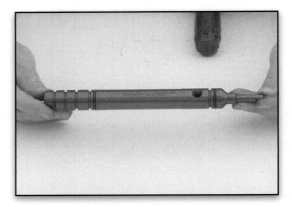

Figure 18.4 JP Rod Guide

Figure 18.3 Bore Snake

Rod Guide: A cleaning rod guide will make life easier during the cleaning process; it will also eliminate some of the mess. The most important aspects of these guides are the fact that you won't bend your rod when cleaning the rifle and the rod has a better chance of sliding smoothly into the bore eliminating damage to the throat of your rifles chamber. Some bore guides are also equipped with a small hole to add solvent without getting it on your hands.

Bore Brushes: If you had to pick only one bore brush to use, it should be a brass bore brush. These brushes work well and will not damage your rifling. You should never use a stainless steel brush on the bore of your rifle because it will cause damage. Plastic bore brushes work well too, but they are not my top choice.

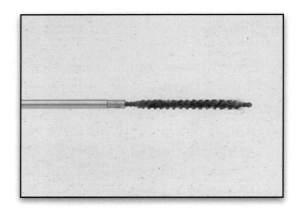

Figure 18.5 Bore Brush

Stab Jag: A stab jag is used to push cleaning patches down the barrel. When I clean a rifle, I always push the rod from the rear of the rifle to the front; this makes cleaning easy and does not cause damage to the crown of the barrel. A stab jag is simply a small, brass part that allows you to stab a patch and push it down the bore.

Most shooters use an eyelet to lace the patch through and then push down the bore. These work fine, but a stab jag gives you more uniform cleaning. If you have a GI rod, you will have to use an eyelet for cleaning.

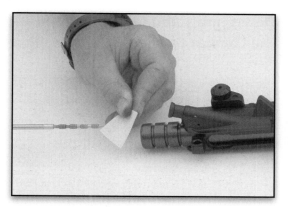

Figure 18.7 Cleaning patches

Copper Solvent: Shooters Choice or Hoppes Number 9 Copper Solvent will get the job done. If you have a special brew or certain brand that you think works better, use it.

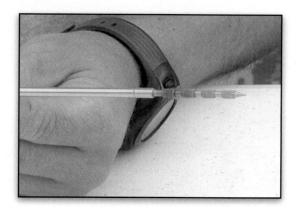

Figure 18.6A Stab Jag

Allen Wrench set: If you attach a scope to your rifle, it will more than likely have allen screws to hold the scope in place. These wrenches come in handy all the time. I prefer the type with a ball end on one end of the wrench. This allows for easily tightening screws that are in awkward positions.

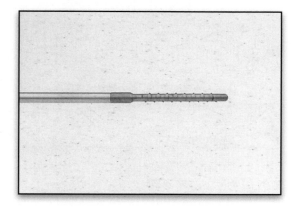

Figure 18.6B Twist Jag

Patches: Get some good patches. Almost anything will work. I use whatever type of patches that I can get my hands on. The military issue square patches are my favorite; they are a little tight on a stab jag but do a great job of getting the bore clean.

Figure 18.8 Allen Wrench Set

Disassembly and Cleaning the Combat Carbine

I have tried to keep cleaning as efficient as possible. I would rather spend time on the range than cleaning, so here is the procedure I follow.

First, I ensure that the weapon is clear. After the weapon is clear, I make sure the bolt is forward and remove the take down pins. Figure 18.10A. This will allow you to separate the upper and lower receivers. Figure 18.10B.

Figure 18.10A Removing take down pins

Figure 18.10B Separating the upper and lower receiver groups

Once the upper and lower receiver are separated, I grab the charging handle and pull slightly to the rear. This will allow you to grasp the bolt carrier and remove it from the weapon.

Figure 18.10C Pull the charging handle slightly to the rear, which will also pull the bolt to the rear.

Figure 18.10D Remove the bolt carrier group from the upper receiver.

Figure 18.10E Pull the charging handle to the rear, once it hits the slot, push the charging handle down and away to realease.

Figure 18.10F Remove the charging handle from the upper receiver.

Figure 18.10H Remove the firing pin retaining pin.

Figure 18.10G Charging handle and carrier group removed and ready for cleaning.

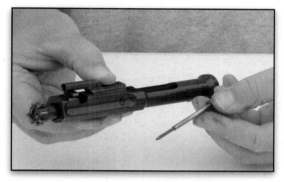

Figure 18.10I Remove firing pin from rear of bolt carrier.

After removing the bolt carrier from the upper, it is time to start the cleaning process. To break the bolt carrier down for cleaning, remove the firing pin retainer pin. Once this pin is removed, you should be able to remove the firing pin by either tapping the rear of the carrier on your bench or by simply tipping the carrier so that gravity will take over.

Once the firing pin is removed, you will need to push the bolt to the rear in the carrier. This will allow you to access the cam pin. The cam pin must be rotated 90 Degrees in order to be removed. Once the cam pin is removed, the bolt can be pulled from the front of the carrier.

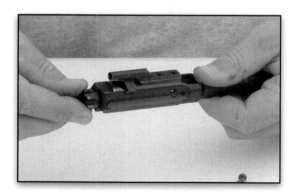

Figure 18.10J Push bolt to the rear, allowing access to the cam pin.

Figure 18.10K Twist cam pin 90 degrees.

Figure 18.10L Remove cam pin.

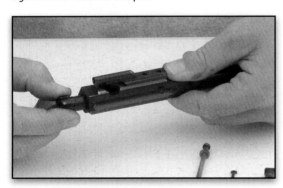

Figure 18.10M Pull bolt from bolt carrier.

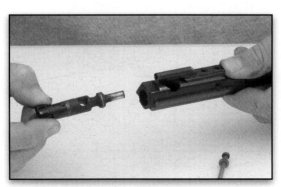

Figure 18.10N Bolt removed from carrier.

After the bolt is removed from the carrier, the extractor must be removed for cleaning. To remove the extractor pin, I apply pressure to the extractor as I push the pin out with the firing pin. Use a punch if you have one, but the firing pin will work.

Figure 18.10O Use firing pin to remove extractor pin, you may need to apply a little pressure on the extractor to remove pin.

Figure 18.10P Extractor pin removed.

Figure 18.10Q Extractor removed.

The next step is to start the cleaning process. First, remove excess carbon from the rear of the bolt with a small screw driver or any object with an edge. Do not use a knife. You will also need to use a weapon's cleaning brush or an old tooth brush to clean the rest of the bolt.

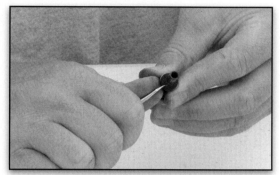

Figure 18.10R Use small screwdriver or sharp metal object to remove carbon from the rear of the bolt.

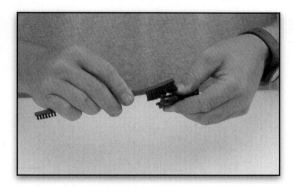

Figure 18.10S Use a weapons cleaning brush or old tooth brush to clean locking lugs on the bolt.

You will also need to clean the rest of the bolt carrier before you reassemble the carrier group. This will require a little patience since there are some areas that are really hard to reach. I use a pistol cleaning rod to get inside of the bolt carrier. I also use small cotton swabs to get in the hard-to-reach spots.

Once the bolt and carrier are cleaned, it is time to reassemble these parts. Start by replacing the extractor and extractor pin in the bolt. Once again, you may need to apply a little pressure to get the pin to snap in.

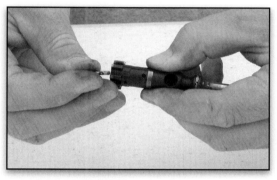

Figure 18.10T Once you have cleaned the entire bolt, replace the extractor and pin.

You will also need to check the gas rings on the bolt before putting the weapon back together. Perform a visual inspection to ensure that the gas rings are not aligned. The weapon will work with the rings aligned, but the ends of the rings can bind on each other as the bolt is pushed into the carrier. If they are aligned, use your fingernail to mis align them. Once the rings are squared away, slide the bolt back into the carrier. Do not force this part; if you are too rough, the gas rings may become damaged.

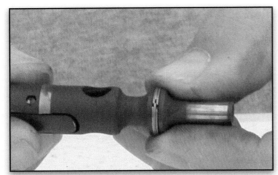

Figure 18.10U Check gas rings to make sure they are not aligned.

Figure 18.10V If gas rings are aligned, use finger nail to move the rings until the gaps to not line up.

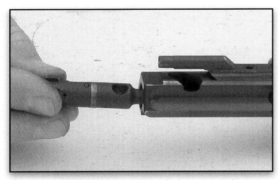

Figure 18.10W Slide bolt back into bolt carrier. Bolt carrier must be cleaned inside and out before replacing bolt.

Place a liberal amount of lube on the bolt before you put the parts back together. Slide the bolt all the way into the carrier; this will allow you to twist the bolt until the cam pin hole is in the correct position. Replace the cam pin and twist it 90 degrees. Pull the bolt forward and slide the firing pin in from the rear of the carrier. Replace the firing pin retainer pin.

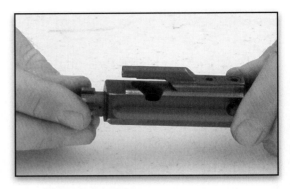

Figure 18.10X Slide bolt all the way into carrier, ensure that the extractor is facing to the right and the hole for the cam pin is aligned with its slot.

Figure 18.10Y Replace cam pin in slot.

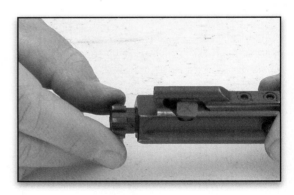

Figure 18.10Z Cam pin in place.

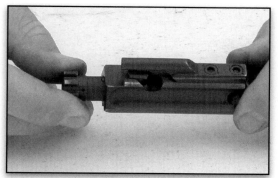

Figure 18.11A Pull bolt forward in carrier.

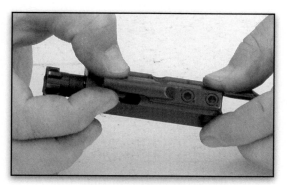

Figure 18.11B Rotate cam pin 90 degrees which will allow firing pin to slide from the rear.

Figure 18.11C Drop firing pin into carrier from the rear.

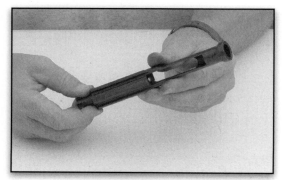

Figure 18.11D Ensure that the firing pin is all the way forward.

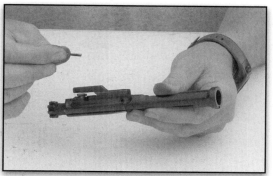

Figure 18.11E Replace firing pin retainer pin.

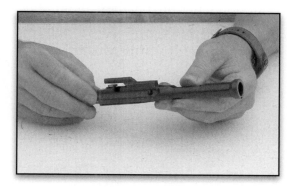

Figure 18.11F Fully assembled bolt carrier group.

Cleaning the upper receiver

Attempt to clean as much of the upper receiver as possible before cleaning the barrel and barrel extension. To date, I haven't found an easy method to do this. You just have to dig in with a little elbow grease. Shop towels, cotton swabs, and gun solvent will get the job done. Once the upper receiver is acceptable, then it is time to get down to business with the barrel.

Start by inserting a rod guide into the upper receiver. This is not a necessity but it does make the process easier. I prefer to use the JP Rod Guide; it has an access point for applying solvent so you don't get it on your hands. Once the guide is in position, use your rod with a stab jag or eyelet and swab the barrel with a wet patch. Once the wet patch has been removed, follow it up with a clean, dry patch. If you allow the barrel to sit for a few minutes with the copper solvent in the barrel, you will see an almost bluish green coloring on the clean patch. This color indicates that you are getting the copper removed from the barrel. I continue to run patches, alternating between wet and dry until there isn't any color coming out on the patches.

Figure 18.11G Install rod guide after you have cleaned the inside of the upper receiver.

Figure 18.11H Push rod guide forward until it is seated.

Figure 18.11I Place patch with copper solvent on the end of your stab jag.

Figure 18.11J Push rod into guide with patch.

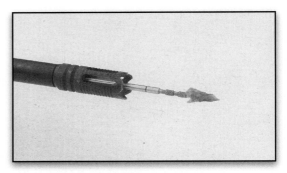

Figure 18.11K Push rod through barrel with even pressure and discard used patch.

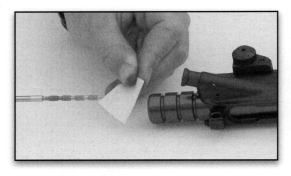

Figure 18.11L Follow wet patch with a clean and dry patch.

Figure 18.11M Stab jag makes cleaning simpler and faster.

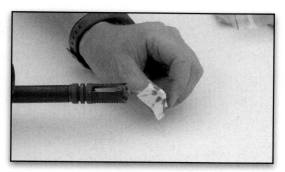

Figure 18.11N Continue to run one wet and one dry until you are not getting any more copper to come out of the barrel.

After completing the barrel cleaning process, you will also need to clean the chamber of the rifle. To clean the chamber, use a chamber brush on a pistol cleaning rod, or on a full length rod. The pistol rod makes this task much easier. Run the chamber brush into the chamber and rotate clockwise. After using the brush, follow it up with a dry patch to get any remaining debris from the chamber.

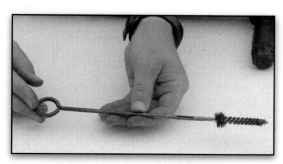

Figure 18.11O Pistol rod with chamber brush.

Figure 18.11P Slide chamber brush into the rear of the receiver.

Figure 18.11Q Rotate chamber brush clockwise in chamber several revolutions.

The last step to clean the upper receiver is to clean the locking lug recesses. A handy product is the Chamber Star from Buffer Technologies. These specially cut swabs make cleaning this area much easier. I generally place one on a bore brush and use it to clean the lug area. A few simple turns and the area is complete.

Figure 18.11R Use small felt swabs to clean locking lug recesses.

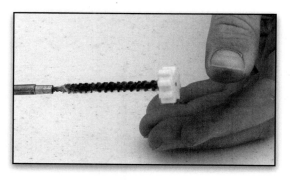

Figure 18.11S Place swab on bore brush.

Figure 18.11T Rotate swab in locking lug recess until area is clean.

The Bore Snake also works well if you need to quickly clean your Carbine's barrel. This nifty device allows the shooter to take a couple minutes and clean their rifle without disassembly. Drop the weighted end into the chamber until it comes through the muzzle; then pull the Bore Snake through the barrel. Repeat until you are happy.

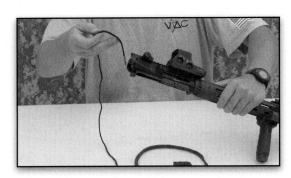

Figure 18.11U If you don't have a cleaning rod the bore snake works well.

Figure 18.11V Drop weighted end into chamber and through barrel.

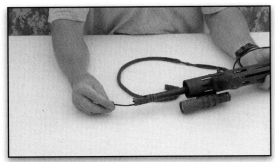

Figure 18.11W Grab free running end and pull.

Figure 18.11X Pull snake through slowly.

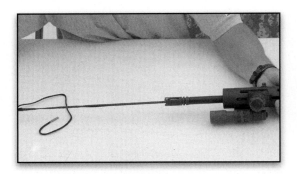

Figure 18.11Y Repeat as necessary.

Your optics should be cleaned during the weapon cleaning process. I use a Lens Pen if possible. This pen lets you choose from a brush or squeegee.

Figure 18.12A Lens Pen.

Figure 18.12B Use lens pen for cleaning optics.

Figure 18.13B Slide mag brush in and out several times.

Figure 18.12C Use either end, both work well.

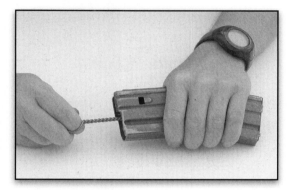

Figure 18.13C Make sure you push the brush all the way to the bottom of the magazine.

Your magazines should be cleaned periodically as well, especially after exposure to a dusty, desert environment. If you have time, disassemble the magazines and clean them thoroughly. If you don't have time, use a magazine brush to allow quick cleaning of your magazines.

Give the lower receiver some attention before reassembling the rifle. A brush should get this job done, unless you have been exposed to a very harsh environment. If this is the case, then you may need to completely disassemble the lower half of the weapon for cleaning.

Figure 18.13A Magazine brush.

Figure 18.14D Clean lower receiver with a weapons brush.

After you have completed the cleaning process, reassemble the weapon to get ready for the fight. As you reassemble the weapon, place additional lube on the bearing surfaces of the bolt carrier, buffer spring, and hammer.

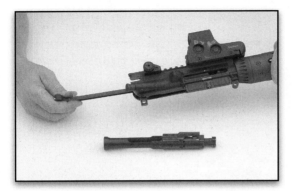

Figure 18.14A Replace charging handle in upper receiver.

Figure 18.14B Oil bolt carrier on all bearing surfaces. Make sure bolt is lubed as well.

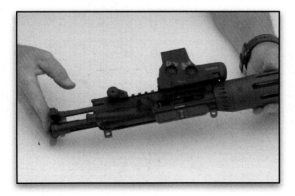

Figure 18.14C Slide bolt carrier back into upper receiver.

Figure 18.14E Place upper and lower back together.

Figure 18.14F Replace takedown pins.

Functions Check

After assembling the weapon, conduct a Functions Check. Follow these procedures: Make sure the weapon is clear. Place the weapon on safe, with the bolt forward, and squeeze the trigger. The hammer should not fall. Place the weapon on fire and squeeze the trigger; the hammer should fall. Keep the trigger squeezed and cycle the bolt. Slowly release the trigger. You should feel and hear the dis-connector click. This procedure is for semi automatic rifles only. Fully automatic rifles have several additional steps.

Over the years I have seen shooters incorrectly assemble their rifles. Most of the deficiencies were due to incorrectly placing the firing pin in the bolt carrier. If the firing pin retainer pin is not installed correctly, the rifle will not work. Double check as you re-assemble the weapon. If there is an opportunity, test fire your weapon.

Springs

The AR has many different springs, some of which will last forever, others that need regular maintenance or the weapon will start to stumble. The springs that require heavy maintenance are the extractor spring and the ejector spring. These springs only cost a few dollars but will make a huge difference. The extractor spring is located under the extractor. There are several aftermarket springs that will improve the function of this part. I use the wolf springs as well as upgrade the rubber stop to the harder, black version.

The ejector spring is a little harder to analyze and to replace. You may need to have a gunsmith fix this problem. If you start to have spent cases fail to leave the ejection port, the ejector could be the problem. You may also notice small brass marks at the front of the ejection port, on the delta ring or on your after market hand guard.

The buffer spring may also need to be changed after several years of abuse. If your weapon seems sluggish, replacing the buffer spring will get the weapon to feel snappy again.

The AR is extremely reliable when it is taken care of. Don't let her down, and she won't let you down.

Figure 18.15A Extractor and extractor spring.

19. TRAINING AIDS

This chapter will give you several things to think about. It isn't meant to sell you on any one item, just to give you a little insight into what may be available to help you become a better tactical shooter. Most sections in this chapter are stand alone. Read what may be of interest and disregard what isn't of interest. If you think some of the statements I will make in this chapter are counter to each other, you are right. Different strokes for different folks, one answer doesn't work for everyone. This is contrary to what some believe, but I have seen it all and there is more than one way to skin a cat. No matter how you do it, you better hang on.

Dummy Rounds

We have already discussed the use of Dummy Rounds. Dummy rounds can be used almost every time you head to the range if you are trying to become a well rounded shooter. These inexpensive training rounds can be used for training shooters to deal with unexpected malfunctions as well as transitions to the sidearm.

When using these dummy rounds, it is important that you ensure that you have training ammunition that is easily recognizable. This will aid in recovering the rounds when they are ejected during malfunction drills. I prefer the inexpensive orange dummy rounds. Also, be absolutely sure that all magazines are inspected for dummy rounds. Completely download the magazines if that is what it takes to be positive.

Green/Frangible Ammunition
Using Green Ammo to gain the Advantage

For the last ten years we have watched environmentalists encroach on the shooting community. Their biggest push has been pressure to eliminate lead in ammunition. However, before that came along, testing was being conducted to try to incorporate frangible types of ammunition into tactical training, simply to allow more pistol safe facilities to be used with rifle calibers. Mild steel can withstand a serious beating from pistol calibers; a rifle, on the other hand, especially 5.56mm, can cause immediate damage to the steel as well as ricochets in the shooter's direction. Even when the best Armor Grade steels are used, 5.56 Green Tip M855 will cause damage. The damage isn't always significant, but over time it will render your steel un-useable or unsafe.

Being a hunter, conservationist, and tactical trainer, I feel the use of Green frangible ammunition is a winner all the way around. From a health standpoint, I already have more than my fair share of build up. This ammo will at least help to prolong my agony, guess this is a good thing?

From a trainer's perspective, this ammo allows more realistic tactical shooting instruction. Not only can you use your normal weapon configuration during close quarters scenarios conducted in live fire facilities, but you can also maximize training at your conventional range.

More often than not, shooters train on paper close, and steel for distant marksmanship training. A large amount of time is wasted pasting targets when shooters could use frangible ammunition on close range steel. Steel is a great training tool when used correctly. It almost eliminates the need for target repair, with the exception of a little paint now and then.

Shooting steel at close range with rifle will help tactical shooters to gain more confidence in their ability. The immediate feedback from a hit or miss will help the shooter better understand corrections that need to be made to make them successful. Paper is a great introductory tool, but the use of steel in close with frangible ammunition will take you to the next level of performance.

Using steel with frangible also allows shooters to share targets without the issue of knowing whose holes are theirs. With frangible on steel the shooter can feel confident he knows where his hits were. You can also use peculiar shaped targets to enhance the training.

Once you have decided to maximize your training time with frangible ammunition and steel, you will make huge leaps in ability. Just remember, this is for training only. Shoot your long targets for zero with normal duty ammunition, don't mix the two types of ammunition ever...

.22 Conversion Kits

When it comes time to head to the range, there are only a few options for close up steel shooting with the AR. One option is to use frangible ammunition. This ammo is great for training, but it will still cost just about as much as reliable tactical ammo.

Another option for training at close quarter range is to use a .22 conversion kit. These kits will not allow you to feel completely as if you were shooting your duty load, but I have not found a more inexpensive way to achieve training goals. Several indoor ranges I have fired at would not let me use the AR due to its high probability of damaging their steel; however, after asking if I could shoot a .22 rifle I met absolutely no resistance.

You can get these kits for a relatively small cost, they will last for years, and the training you will achieve by shooting more ammunition through your weapon is priceless.

9MM Upper Receivers

The 9mm upper will offer you similar advantages to the .22 conversion, as well as the use of frangible ammunition. Down falls of using frangible are the expense and the damage it will causes to your steel. A disadvantage when using the .22 conversion is the fact that you will not experience the recoil associated with full up, self defense ammunition. 9mm receivers are available from several manufacturers and there are also numerous places to buy 9mm in bulk relatively cheap; at least cheaper than frangible ammo. You are also able to set up the 9mm upper to feel similar to the normal upper that you will carry on your rifle. Numerous indoor ranges have banned the firing of 5.56mm ammunition. This upper will let you continue your training, regardless of the ranges no 5.56mm rule. Lastly, there may be a time when using the 9mm upper for some type of scenario makes sense. It is just another option to accomplish the training you need to be successful in the field.

Paint Marking Weapons

Weapons used for force on force training have created a completely new venue to train the tactical shooter. These tools have evolved and will continue to evolve in the future. To date, there has not been a reliable, realistic, paint marking weapon produced that closely simulates firing a real weapon. The realism isn't of the utmost importance. What is important is how well trainers can implement a training program that simulates real life.

If instructors let tactical training evolve into paint ball wars, there will be no significant gain in capability from this training. Real bullets will have devastating effects on not only the enemy, but on you and your team as well. Real bullets will go through walls, so if bad guys paint a wall opposite of your position, the trainer must treat the situation as real as can be, without firing actual bullets into real people. Maybe we will someday be able to use criminals or terrorists for target practice, until then we need to monitor paint gun training with a prudent eye.

Several instructors have designed facilities that simulate real buildings when used with paint marking rounds. They have used such techniques as paper strung near doorways which enables the bad guy to shoot through the walls of the facility, as it would be with real bullets in real sheetrock walls.

These instructors have also spent a lot of time supervising and coaching their role players to act as they should when engaged by the officers who are conducting the training event. If your role players are just there for paint ball wars, they will not help the training reach a realistic level. Find role players that understand the importance of playing by the rules, it will increase your survivability in an actual gunfight. ***Training is training, not a game.***

Air Soft Weapons

Air soft weapons have been around for years, but, until recently, the quality and realism left a lot to be desired. Today, you can find plenty of reliable options in the air soft world. Air soft weapons have been in use in the Far East for years for gaming purpose. In the US we have scoffed at the use of air weapons to train tactical shooters, it just doesn't seem macho enough, I guess. These weapons can be used to not only practice your techniques without a live fire range, but they also allow you to conduct marksmanship training in your house or garage. If you want to conduct team tactical training, the use of air soft weapons will open up more opportunities for your team. These weapons will allow use of facilities that wouldn't be available for paint or bullet trap use. You will not have to deal with the extensive set up or clean up time needed for paint or live fire into bullet traps.

Training Guns - Blue/Red Guns

Blue and Red guns are plastic training weapons that have similar weight and feel to normal duty weapons. These weapons can be rigged with slings, lights, and sighting systems if necessary. Blue and Red guns provide shooters with the chance to practice positions, retention drills, or demonstration without possible damage to live weapons, or unsafe conditions with other trainees. Live weapons should not be used when conducting drills that involve pointing weapons at other shooters. This is where training weapons add benefit without much cost.

Visible Lasers for Training

Visible lasers are definitely applicable to tactical operations. They are also a handy tool for flat range training as well as dry fire. The visible laser can be used to determine shooter error for such drills as shooting on the move. If the laser is bouncing on the target, the shooter will be able to analyze his errors and adjust as needed to achieve smooth foot movement and better accuracy on the target.

Don't become over dependant on the use of the laser. Whether for training or tactical scenarios, you must be able to handle the situation without the laser. Sometimes Mr. Murphy doesn't like us to be able to employ these great tools.